対訳

# IATF 16949:2016

ポケット版

## 自動車産業品質マネジメントシステム規格——自動車産業の生産部品及び関連するサービス部品の組織に対する品質マネジメントシステム要求事項

日本規格協会　編

International
Automotive
Task Force

日本規格協会

# ご利用にあたって

IATF 16949:2016 は，ISO 9001 に自動車産業固有の要求事項を加えたセクター規格である ISO/TS 16949:2009 をその前身としており，ISO 9001 の 2015 年改訂に対応する際，ISO 規格ではなく IATF 規格として発行されたものである．

ISO/TS 16949:2009 ではその本文に ISO 9001:2008 の箇条を要求事項として引用していたが，IATF 16949:2016 では ISO 9001:2015 の箇条番号及び題名のみを引用し，本文は ISO 9001:2015 を参照する構成となっている．IATF 16949:2016 は，単独の品質マネジメントシステム規格とはみなされず，ISO 9001:2015 の補足として使用することが求められていることからも，本書読者におかれては，姉妹書『対訳 ISO 9001:2015 (JIS Q 9001:2015) 品質マネジメントの国際規格［ポケット版］』と併せて使用されることをお勧めする．

本書収録の日本語訳は，AIAG(Automotive Industry Action Group)が作成したものであるが，日本語訳に疑義があるときは規格原文に準拠されたい．日本語訳のみを使用して生じた不都合な事態に関して，当会及び AIAG は一切責任を負わない．原文のみが有効である．

2016 年 12 月

日本規格協会

# IATF 16949

First edition 2016-10-01

## Automotive Quality Management System Standard —

Quality management system requirements for automotive production and relevant service parts organizations

## 自動車産業品質マネジメントシステム規格—

自動車産業の生産部品及び関連するサービス部品の組織に対する品質マネジメントシステム要求事項

## IATF copyright notice



Requests for permission to reproduce and/or translate any part of this Automotive QMS Standard should be addressed to one of the following national automotive trade associations below:

Associazione Nazionale Filiera Industrie Automobilistiche (ANFIA / Italy)

Automotive Industry Action Group (AIAG / USA)

Fédération des Industries des Équipements pour Véhicules (FIEV / France)

Society of Motor Manufacturers and Traders Ltd. (SMMT / UK)

Verband der Automobilindustrie e.V. (VDA / Germany)

## IATF 著作権表示

# Contents

# 目　　次

ページ

## Foreword — Automotive QMS Standard

This Automotive Quality Management System Standard, herein referred to as "Automotive QMS Standard" or "IATF 16949," along with applicable automotive customer-specific requirements, ISO 9001:2015 requirements, and ISO 9000:2015 defines the fundamental quality management system requirements for automotive production and relevant service parts organizations. As such, this Automotive QMS Standard cannot be considered a stand-alone QMS Standard but has to be comprehended as a supplement to and used in conjunction with ISO 9001:2015. ISO 9001:2015 is published as a separate ISO Standard.

IATF 16949:2016 (1st edition) represents an innovative document, given the strong orientation to the customer, with inclusion of a number of consolidated previous customer specific requirements.

Annex B is provided for guidance to implement the IATF 16949 requirements unless otherwise specified by customer specific requirements.

## まえがき―自動車産業 QMS 規格

この自動車産業品質マネジメントシステム規格は，ここでは“自動車産業 QMS 規格”又は“IATF 16949”と呼び，該当する自動車産業の顧客固有要求事項，ISO 9001:2015 要求事項及び ISO 9000:2015 とともに，自動車産業の生産部品及び関連するサービス部品の組織に対する基本的品質マネジメントシステム要求事項を定める．よって，この自動車産業 QMS 規格は，単独の QMS 規格とはみなされず，ISO 9001:2015 の補足として理解しなければならない．また，ISO 9001:2015 と併せて使用しなければならない．ISO 9001:2015 は，別個の ISO 規格として発行されている．

IATF 16949:2016（第 1 版）は，以前の顧客固有要求事項を取り入れてまとめた，強い顧客志向を与えられた，革新的文書を示している．

附属書 B は，IATF 16949 要求事項を実施する手引として提供されている．ただし，顧客固有要求事項によって規定されたときは，この限りではない．

## History

ISO/TS 16949 (1st edition) was originally created in 1999 by the International Automotive Task Force (IATF) with the aim of harmonizing the different assessment and certification systems worldwide in the supply chain for the automotive sector. Other revisions were created (2nd edition in 2002, and 3rd edition in 2009) as necessary for either automotive sector enhancements or ISO 9001 revisions. ISO/TS 16949 (along with supporting technical publications developed by original equipment manufacturers [herein referred to as OEMs] and the national automotive trade associations) introduced a common set of techniques and methods for common product and process development for automotive manufacturing worldwide.

In preparation for migrating from ISO/TS 16949: 2009 (3rd edition) to this Automotive QMS Standard, IATF 16949, feedback was solicited from certification bodies, auditors, suppliers, and OEMs to create IATF 16949:2016 (1st edition), which cancels and replaces ISO/TS 16949:2009

## 歴史

ISO/TS 16949（第1版）は，自動車産業界のサプライチェーンにおける異なる評価及び認証システムを調和させる目的で，1999年，International Automotive Task Force（IATF）によって最初に作成された．他の改訂（2002年第2版，2009年第3版）は，自動車産業界の強化又はISO 9001の改訂に対する必要に応じて，作成された．ISO/TS 16949［自動車メーカー（ここではOEMと呼ぶ）及び国家自動車産業団体によって開発された裏付けとなる技術的出版物とともに］は，世界的な自動車産業の製造のために，共通の製品及び工程の開発のための一連の手法及び方法を導入した．

ISO/TS 16949:2009（第3版）から，この自動車産業QMS規格IATF 16949に移行する準備において，ISO/TS 16949:2009（第3版）を失効させ置き換えるIATF 16949:2016（第1版）を開発するために，認証機関，審査員，供給者及びOEMからフィードバックが求められた．

(3rd edition).

The IATF maintains strong cooperation with ISO by continuing liaison committee status ensuring continued alignment with ISO 9001.

## Goal

The goal of this Automotive QMS standard is the development of a quality management system that provides for continual improvement, emphasizing defect prevention and the reduction of variation and waste in the supply chain.

## Remarks for certification

Requirements for certification to this Automotive QMS Standard are defined in the Rules for achieving and maintaining IATF recognition.

Details can be obtained from the local Oversight Offices of the IATF cited below:

Associazione Nazionale Filiera Industrie Au-

IATF は，ISO 9001 との継続的整合を確実にするリエゾン委員会の地位を継続することによって，ISO との強固な協力を維持している．

## 到達目標

この自動車産業 QMS 規格の到達目標は，不具合の予防，並びにサプライチェーンにおけるばらつき及び無駄の削減を強調した，継続的改善をもたらす品質マネジメントシステムを開発することである．

## 認証に対する注意点

この自動車産業 QMS 規格への認証に対する要求事項は，IATF 承認取得及び維持のためのルールに定められている．

詳細は，下記に引用されている，地域の IATF 監督機関から入手できる．

Associazione Nazionale Filiera Industrie Au-

tomobilistiche (ANFIA)
Web site: www.anfia.it
e-mail: anfia@anfia.it

International Automotive Oversight Bureau (IAOB)
Web site: www.iaob.org
e-mail: iatf16949feedback@iaob.org

IATF France
Web site: www.iatf-france.com
e-mail: iatf@iatf-france.com

Society of Motor Manufacturers and Traders Ltd. (SMMT Ltd.)
Web site: www.smmtoversight.co.uk
e-mail: iatf16949@smmt.co.uk

Verband der Automobilindustrie – Qualitäts Management Center (VDA QMC)
Web site: www.vda-qmc.de
e-mail: info@vda-qmc.de

All public information about the IATF can be

tomobilistiche (ANFIA)
Web site: www.anfia.it
e-mail: anfia@anfia.it

International Automotive Oversight Bureau (IAOB)
Web site: www.iaob.org
e-mail: iatf16949feedback@iaob.org

IATF France
Web site: www.iatf-france.com
e-mail: iatf@iatf-france.com

Society of Motor Manufacturers and Traders Ltd. (SMMT Ltd.)
Web site: www.smmtoversight.co.uk
e-mail: iatf16949@smmt.co.uk

Verband der Automobilindustrie – Qualitäts Management Center (VDA QMC)
Web site: www.vda-qmc.de
e-mail: info@vda-qmc.de

IATF に関する全ての公開情報は，IATF ウェブ

found at the IATF website:
www.iatfglobaloversight.org

サイト www.iatfglobaloversight.org で公表されている.

# Introduction

## 0.1 General

See ISO 9001:2015 requirements.

## 0.2 Quality management principles

See ISO 9001:2015 requirements.

## 0.3 Process approach

### 0.3.1 General

See ISO 9001:2015 requirements.

### 0.3.2 Plan-Do-Check-Act cycle

See ISO 9001:2015 requirements.

### 0.3.3 Risk-based thinking

See ISO 9001:2015 requirements.

## 0.4 Relationship with other management system standards

See ISO 9001:2015 requirements.

# Quality management systems — Requirements

## 序文

### 0.1　一般

ISO 9001:2015 要求事項参照.

### 0.2　品質マネジメントの原則

ISO 9001:2015 要求事項参照.

### 0.3　プロセスアプローチ

#### 0.3.1　一般

ISO 9001:2015 要求事項参照.

#### 0.3.2　PDCA サイクル

ISO 9001:2015 要求事項参照.

#### 0.3.3　リスクに基づく考え方

ISO 9001:2015 要求事項参照.

### 0.4　他のマネジメントシステム規格との関係

ISO 9001:2015 要求事項参照.

## 品質マネジメントシステム—要求事項

## 1 Scope

See ISO 9001:2015 requirements.

### 1.1 Scope — automotive supplemental to ISO 9001:2015

This Automotive QMS Standard defines the quality management system requirements for the design and development, production and, when relevant, assembly, installation, and services of automotive-related products, including products with embedded software.

This Automotive QMS Standard is applicable to sites of the organization where manufacturing of customer-specified production parts, service parts, and/or accessory parts occur.

This Automotive QMS Standard should be applied throughout the automotive supply chain.

## 2 Normative references

See ISO 9001:2015 requirements.

## 1　適用範囲

ISO 9001:2015 要求事項参照.

### 1.1　適用範囲—ISO 9001:2015 に対する自動車産業補足

この自動車産業 QMS 規格は，組込みソフトウェアをもつ製品を含む，自動車関係製品の，設計・開発，生産，該当する場合，組立，取付け及びサービスの品質マネジメントシステム要求事項を定める.

この自動車産業 QMS 規格は，顧客規定生産部品，サービス部品及び／又はアクセサリー部品の製造を行う組織のサイトに適用する.

この自動車産業 QMS 規格は，自動車産業サプライチェーン全体にわたって適用することが望ましい.

## 2　引用規格

ISO 9001:2015 要求事項参照.

## 2.1 Normative and informative references

Annex A (Control Plan) is a normative part of this Automotive QMS standard.

Annex B (Bibliography — automotive supplemental) is informative, which provides additional information intended to assist the understanding or use of this Automotive QMS standard.

## 3 Terms and definitions

See ISO 9001:2015 requirements.

## 3.1 Terms and definitions for the automotive industry

**accessory part**

customer-specified additional component(s) that are either mechanically or electronically connected to the vehicle or powertrain before (or after) delivery to the final customer (e.g., custom floor mats, truck bed liners, wheel covers, sound system enhancements, sunroofs, spoilers, super-chargers, etc.)

## 2.1　規定及び参考の引用

附属書 A（コントロールプラン）は，この自動車産業 QMS 規格の規定部分である.

附属書 B（参考文献—自動車産業補足）は，参考であり，この自動車産業 QMS 規格の理解又は使用の支援になることを意図した追加情報を提供するものである.

> **3　用語及び定義**
>
> ISO 9001:2015 要求事項参照.

## 3.1　自動車産業の用語及び定義

**アクセサリー部品（accessory part）**

最終顧客への引渡しの事前（又は事後）に，機械的又は電子的に車両又はパワートレーンに結合される，顧客規定の追加構成部品（例　特注フロアマット，トラックベッドライナー，ホイールカバー，音響システム機能強化装置，サンルーフ，スポイラー，スーパーチャージャーなど）.

**advanced product quality planning (APQP)**

product quality planning process that supports development of a product or service that will satisfy customer requirements; APQP serves as a guide in the development process and also a standard way to share results between organizations and their customers; APQP covers design robustness, design testing and specification compliance, production process design, quality inspection standards, process capability, production capacity, product packaging, product testing and operator training plan, among other items

**aftermarket part**

replacement part(s) not procured or released by an OEM for service part applications, which may or may not be produced to original equipment specifications

**authorization**

documented permission for a person(s) specifying rights and responsibilities related to giving or denying permissions or sanctions within an organi-

**先行製品品質計画［advanced product quality planning (APQP)］**

顧客要求事項を満たす製品又はサービスの開発を支援する，製品品質計画プロセス．APQP は，開発プロセスにおいて手引として活用し，組織と顧客との間で結果を共有化する標準的方法でもある．APQP は，数ある中でも特に，設計の頑健性，設計試験及び仕様への適合，生産工程設計，品質検査規格，工程能力，生産能力，製品の包装，製品試験及び作業者教育訓練計画を網羅している．

**アフターマーケット部品（aftermarket part）**

サービス部品用途として OEM が調達又はリリースするものではない交換部品で，OEM の仕様どおり製造されてもそうでなくてもよい．

**正式許可（authorization）**

組織内で，許可を与える若しくは拒否する，又は制裁することに関係する権限及び責任を規定する人に対する，文書化した許可．

zation

**challenge (master) part**

part(s) of known specification, calibrated and traceable to standards, with expected results (pass or fail) that are used to validate the functionality of an error-proofing device or check fixtures (e.g., go / no-go gauging)

**control plan**

documented description of the systems and processes required for controlling the manufacturing of product (see Annex A)

**customer requirements**

all requirements specified by the customer (e.g., technical, commercial, product and manufacturing process-related requirements, general terms and conditions, customer-specific requirements, etc.)

**customer-specific requirements (CSRs)**

**チャレンジ（マスター）部品［challenge (master) part］**

ポカヨケ装置の機能又は点検ジグ（例　通止ゲージ）の妥当性確認に使用する，既知の仕様，校正された及び標準にトレーサブルな，期待された結果（合格又は不合格）をもつ部品．

**コントロールプラン（control plan）**

製品の製造を管理するために要求される，システム及びプロセスを記述した文書（附属書A参照）．

**顧客要求事項（customer requirements）**

顧客に規定された全ての要求事項（例　技術，商流，製品及び製造工程に関係する要求事項，一般の契約条件，顧客固有要求事項など）．

**顧客固有要求事項［customer-specific requirements (CSRs)］**

interpretations of or supplemental requirements linked to a specific clause(s) of this Automotive QMS Standard

**design for assembly (DFA)**

process by which products are designed with ease of assembly considerations. (e.g., if a product contains fewer parts it will take less time to assemble, thereby reducing assembly costs)

**design for manufacturing (DFM)**

integration of product design and process planning to design a product that is easily and economically manufactured

**design for manufacturing and assembly (DFMA)**

combination of two methodologies: Design for Manufacture (DFM), which is the process of optimizing the design to be easier to produce, have higher throughput, and improved quality; and Design for Assembly (DFA), which is the optimization of the design to reduce risk of error, lowering costs, and making it easier to assemble

この自動車産業 QMS 規格の特定の箇条にリンクした解釈又は補足の要求事項.

**組立設計［design for assembly (DFA)］**

組立容易性を考慮した製品設計のプロセス（例 製品がより少ない部品点数から成れば，組立時間を少なくでき, したがって組立コストを削減できる).

**製造設計［design for manufacturing (DFM)］**

容易に，かつ，経済的に製造される製品を設計するための製品設計及び工程計画の統合化.

**製造及び組立設計［design for manufacturing and assembly (DFMA)］**

二つの方法論の組合せ．製造設計（DFM）は，より容易に生産するための設計を最適化するプロセスであり，より高いスループット，改善した品質をもつ．組立設計（DFA）は，不具合のリスクを低減する，コストを下げる及び組立しやすくするための設計の最適化である.

**design for six sigma (DFSS)**

systematic methodology, tools, and techniques with the aim of being a robust design of products or processes that meets customer expectations and can be produced at a six sigma quality level

**design-responsible organization**

organization with authority to establish a new, or change an existing, product specification

NOTE This responsibility includes testing and verification of design performance within the customer's specified application.

**error proofing**

product and manufacturing process design and development to prevent manufacture of nonconforming products

**escalation process**

process used to highlight or flag certain issues within an organization so that the appropriate

**シックスシグマ設計［design for six sigma (DFSS)］**

顧客の期待を満たしシックスシグマ品質レベルで生産可能な製品又は工程の頑健な設計を狙いとする，体系的方法論，ツール及び手法．

**設計責任のある組織（design responsible organization）**

新規の製品仕様を確立する，又は既存の製品仕様を変更する権限をもつ組織．

注記　この責任は，顧客の規定した適用条件の中で，設計性能について試験及び検証することを含む．

**ポカヨケ（error proofing）**

不適合製品の製造を予防するための，製品及び製造工程の設計・開発．

**上申プロセス（escalation process）**

組織内のある問題に対して，適切な要員がその状況に対応し，解決策を監視できるよう，その問題を

personnel can respond to these situations and monitor the resolutions

**fault tree analysis (FTA)**

deductive failure analysis methodology in which an undesired state of a system is analysed; fault tree analysis maps the relationship between faults, subsystems, and redundant design elements by creating a logic diagram of the overall system

**laboratory**

facility for inspection, test, or calibration that may include but is not limited to the following: chemical, metallurgical, dimensional, physical, electrical, or reliability testing

**laboratory scope**

controlled document containing

- specific tests, evaluations, and calibrations that a laboratory is qualified to perform;
- a list of the equipment that the laboratory uses to perform the above; and
- a list of methods and standards to which the

指摘する又は提起するために用いられるプロセス.

**故障の木解析［fault tree analysis（FTA）］**

システムの望ましくない状態を解析する演繹故障解析の方法論. 故障の木解析は, システム全体の論理図を創出することによって, 故障, サブシステム及び冗長設計要素との関係を描く.

**試験所（laboratory）**

化学, 金属, 寸法, 物理, 電気又は信頼性の試験を含めてよいが, それに限定されない, 検査, 試験又は校正の施設.

**試験所適用範囲（laboratory scope）**

次の事項を含む管理文書.

- 試験所が実行するために適格性確認された, 特定の試験, 評価及び校正
- 上記を実行するために用いる設備のリスト
- 上記を実行する方法及び規格のリスト

laboratory performs the above

**manufacturing**

process of making or fabricating

- production materials;
- production parts or service parts;
- assemblies; or
- heat treating, welding, painting, plating, or other finishing services

**manufacturing feasibility**

an analysis and evaluation of a proposed project to determine if it is technically feasible to manufacture the product to meet customer requirements. This includes but is not limited to the following (as applicable): within the estimated costs, and if the necessary resources, facilities, tooling, capacity, software, and personnel with required skills, including support functions, are or are planned to be available

**manufacturing services**

companies that test, manufacture, distribute, and

**製造（manufacturing）**

次に示すものを製作する，又は仕上げるプロセス．

- 生産材料
- 生産部品若しくはサービス部品
- 組立製品
- 熱処理，溶接，塗装，鍍金，若しくは他の仕上げ作業

**製造フィージビリティ（manufacturing feasibility）**

製品を，顧客要求事項を満たすように製造することが技術的に実現可能か否かを判定するための，提案されたプロジェクトの分析及び評価．これには，次の事項（該当する場合には，必ず）を含む．しかし，それに限定されない：見積りコスト内で，並びに必要な資源，施設，治工具，生産能力，ソフトウェア及び必要な技能をもつ要員が，支援部門を含めて，提供できるか又は提供できるように計画されているかどうか．

**製造サービス（manufacturing services）**

構成部品及び組立製品に対して，試験，製造，配

provide repair services for components and assemblies

**multi-disciplinary approach**

method to capture input from all interested parties who may influence how a process is administered by a team whose members include personnel from the organization and may include customer and supplier representatives; team members may be internal or external to the organization; either existing teams or ad hoc teams may be used as circumstances warrant; input to the team may include both organization and customer inputs

**no trouble found (NTF)**

designation applied to a part replaced during a service event that, when analysed by the vehicle or parts manufacturer, meets all the requirements of a "good part" (also referred to as "No Fault Found" or "Trouble Not Found")

**outsourced process**

portion of an organization's function (or processes)

給及び修理サービス提供を行う会社.

**部門横断的アプローチ（multi-disciplinary approach）**

メンバーが組織からの要員並びに顧客及び供給者の代表を含めてもよいチームによって，プロセスがどのように管理されるかに影響を及ぼす可能性がある全ての利害関係者からインプットを得る方法．チームメンバーは組織の内部又は外部の要員でもよい．既存のチーム又は事情によっては臨時のチームを使用してもよい．チームへのインプットは組織及び顧客の両方のインプットを含めてもよい.

**不具合なし［no trouble found（NTF）］**

サービス案件が発生したときに交換され，車両又は部品の製造業者によって分析された際に，全ての“良品”の要求事項を満たす部品に適用される呼称［“故障なし（No Fault Found)”又は“不具合発見なし（Trouble Not Found)”とも呼ばれる.］.

**アウトソースしたプロセス（outsourced process）**

外部組織によって実行される，組織の機能（又は

that is performed by an external organization

**periodic overhaul**

maintenance methodology to prevent a major unplanned breakdown where, based on fault or interruption history, a piece of equipment, or subsystem of the equipment, is proactively taken out of service and disassembled, repaired, parts replaced, reassembled, and then returned to service

**predictive maintenance**

an approach and set of techniques to evaluate the condition of in-service equipment by performing periodic or continuous monitoring of equipment conditions, in order to predict when maintenance should be performed

**premium freight**

extra costs or charges incurred in addition to contracted delivery

NOTE This can be caused by method, quantity, unscheduled or late deliveries, etc.

プロセス）の一部.

**定期的オーバーホール（periodic overhaul）**

重大な予期しない故障を予防するために，故障又は中断の経緯に基づいて，設備の一部，又は設備のサブシステムは事前に操業を中止し，分解し，修理し，部品交換し，再度組み立て，再び操業に戻す，保全の方法論.

**予知保全（predictive maintenance）**

いつ保全を実行すべきかを予測するために，定期的又は継続的に設備条件を監視することによって，使用している設備の条件を評価する，方法及び一連の手法.

**特別輸送費（premium freight）**

契約した輸送費に対する割増しの費用又は負担.

注記　これは，輸送方法，量，予定外納入又は納入遅延，などによって発生し得る.

**preventive maintenance**

planned activities at regular intervals (time-based, periodic inspection, and overhaul) to eliminate causes of equipment failure and unscheduled interruptions to production, as an output of the manufacturing process design

**product**

applies to any intended output resulting from the product realization process

**product safety**

standards relating to the design and manufacturing of products to ensure they do not represent harm or hazards to customers

**production shutdown**

condition where manufacturing processes are idle; time span may be a few hours to a few months

**reaction plan**

action or series of steps prescribed in a control plan in the event abnormal or nonconforming events are detected

**予防保全（preventive maintenance）**

製造工程設計のアウトプットとして，設備故障及び予定外の生産中断の原因を除去するために，一定の間隔（時間ベース，定期的検査及びオーバーホール）で計画した活動．

**製品（product）**

製品実現プロセスの結果として生じる，意図したアウトプット．

**製品安全（product safety）**

顧客に危害，危険を与えないことを確実にする，製品の設計及び製造に関係する規範．

**生産シャットダウン（production shutdown）**

製造工程が稼働していない状況．期間は，数時間から数か月でもよい．

**対応計画（reaction plan）**

異常又は不適合事象が検出された場合に，コントロールプランに規定された処置又は一連のステップ．

**remote location**

location that supports manufacturing sites and at which non-production processes occur

**service part**

replacement part(s) manufactured to OEM specifications that are procured or released by the OEM for service part applications, including remanufactured parts

**site**

location at which value-added manufacturing processes occur

**special characteristic**

classification of a product characteristic or manufacturing process parameter that can affect safety or compliance with regulations, fit, function, performance, requirements, or subsequent processing of product

**special status**

notification of a customer-identified classification assigned to an organization where one or more

**遠隔地支援事業所（remote location）**

製造サイトを支援する，生産プロセスのない事業所.

**サービス部品（service part）**

サービス部品としてOEMが調達又はリリースし，OEMの仕様どおり製造される，再生部品を含む交換部品.

**サイト（site）**

価値を付加する製造工程のある事業所.

**特殊特性（special characteristics）**

安全若しくは規制への適合，取付け時の合い，機能，性能，要求事項，又は製品の後加工に影響し得る，製品特性又は製造工程パラメータの区分.

**特別状態（special status）**

著しい品質又は納入問題による一つ以上の顧客要求事項が満たされていない場合に組織に課せられ

customer requirements are not being satisfied due to a significant quality or delivery issue

**support function**

non-production activity (conducted on site or at a remote location) that supports one (or more) manufacturing sites of the same organization

**total productive maintenance**

a system of maintaining and improving the integrity of production and quality systems through machines, equipment, processes, and employees that add value to the organization

**trade-off curves**

tool to understand and communicate the relationship of various design characteristics of a product to each other; a product's performance on one characteristic is mapped on the Y-axis and another on the x-axis, then a curve is plotted to illustrate product performance relative to the two characteristics

**trade-off process**

る，顧客が特定した区分の通知.

**支援機能（support function）**

同じ組織の一つ（又はそれ以上）の製造サイトを支援する，非生産活動（サイト内で又は遠隔地支援事業所で行われる.）.

**TPM（total productive maintenance）**

生産及び品質システムの完全に整った状態を，組織に価値を付加する，機械，設備，工程及び従業員を通じて，維持し改善するシステム.

**トレードオフ曲線（trade-off curves）**

製品の様々な設計特性の相互の関係を理解し伝達するためのツール．一つの特性に関する製品の性能を縦軸に描き，もう一つの特性を横軸に描く．それから二つの特性に対する製品性能を示すために曲線がプロットされる.

**トレードオフプロセス（trade-off process）**

methodology of developing and using trade-off curves for products and their performance characteristics that establish the customer, technical, and economic relationship between design alternatives

## 4 Context of the organization

### 4.1 Understanding the organization and its context

See ISO 9001:2015 requirements.

### 4.2 Understanding the needs and expectations of interested parties

See ISO 9001:2015 requirements.

### 4.3 Determining the scope of the quality management system

See ISO 9001:2015 requirements.

#### 4.3.1 Determining the scope of the quality management system — supplemental

Supporting functions, whether on-site or remote (such as design centres, corporate headquarters, and distribution centres), shall be included in the

製品及びその性能特性に対して，設計の代替案の間で顧客，技術及び経済的な関係を確立する，トレードオフ曲線を作成し使用する方法論.

## 4　組織の状況

### 4.1　組織及びその状況の理解

ISO 9001:2015 要求事項参照.

### 4.2　利害関係者のニーズ及び期待の理解

ISO 9001:2015 要求事項参照.

### 4.3　品質マネジメントシステムの適用範囲の決定

ISO 9001:2015 要求事項参照.

#### 4.3.1　品質マネジメントシステムの適用範囲の決定—補足

支援部門（設計センター，本社及び配給センターのような）は，サイト内にあろうと遠隔地にあろうと，品質マネジメントシステム（QMS）の適用範

scope of the Quality Management System (QMS).

The only permitted exclusion for this Automotive QMS Standard relates to the product design and development requirements within ISO 9001, Section 8.3. The exclusion shall be justified and maintained as documented information (see ISO 9001, Section 7.5).

Permitted exclusions do not include manufacturing process design.

### 4.3.2 Customer-specific requirements

Customer-specific requirements shall be evaluated and included in the scope of the organization's quality management system.

## 4.4 Quality management system and its processes

### 4.4.1

See ISO 9001:2015 requirements.

#### 4.4.1.1 Conformance of products and processes

囲に含めなければならない.

この自動車産業QMS規格における唯一の許可される除外は, ISO 9001の8.3における製品の設計・開発の要求事項に関連するものである. 除外は, 文書化した情報（ISO 9001の7.5参照）として正当性を示し維持しなければならない.

許可される除外に, 製造工程設計は含めない.

#### 4.3.2　顧客固有要求事項

顧客固有要求事項は, 評価し, 組織の品質マネジメントシステムの適用範囲に含めなければならない.

### 4.4　品質マネジメントシステム及びそのプロセス

#### 4.4.1

ISO 9001:2015要求事項参照.

##### 4.4.1.1　製品及びプロセスの適合

The organization shall ensure conformance of all products and processes, including service parts and those that are outsourced, to all applicable customer, statutory, and regulatory requirements (see Section 8.4.2.2).

### 4.4.1.2 Product safety

The organization shall have documented processes for the management of product-safety related products and manufacturing processes, which shall include but not be limited to the following, where applicable:

a) identification by the organization of statutory and regulatory product-safety requirements;
b) customer notification of requirements in item a);
c) special approvals for design FMEA;
d) identification of product safety-related characteristics;
e) identification and controls of safety-related characteristics of product and at the point of manufacture;
f) special approval of control plans and process FMEAs;

組織は，全ての製品及びプロセスが，サービス部品及びアウトソースしたものを含めて，該当する，全ての顧客，法令，規制の要求事項（8.4.2.2 参照）に適合することを確実にしなければならない．

### 4.4.1.2 製品安全

組織は，製品安全に関係する製品及び製造工程の運用管理に対する文書化したプロセスをもたなければならない．それには，該当する場合には，必ず，次の事項を含めなければならない．しかし，それに限定されない．

a) 法令・規制の製品安全要求事項の組織による特定
b) a)における要求事項の顧客からの通知
c) 設計 FMEA に対する特別承認
d) 製品安全に関係する特性の特定
e) 安全に関係する製品特性及び製造時点での特性の特定及び管理
f) コントロールプラン及び工程 FMEA の特別承認

g) reaction plans (see Section 9.1.1.1);

h) defined responsibilities, definition of escalation process and flow of information, including top management, and customer notification;

i) training identified by the organization or customer for personnel involved in product-safety related products and associated manufacturing processes;

j) changes of product or process shall be approved prior to implementation, including evaluation of potential effects on product safety from process and product changes (see ISO 9001, Section 8.3.6);

k) transfer of requirements with regard to product safety throughout the supply chain, including customer-designated sources (see Section 8.4.3.1);

l) product traceability by manufactured lot (at a minimum) throughout the supply chain (see Section 8.5.2.1);

m) lessons learned for new product introduction.

NOTE: Special approval is an additional approv-

g) 対応計画（9.1.1.1参照）
h) 定められた責任，トップマネジメントを含めた上申プロセス及び情報フローの明確化，並びに顧客への通知

i) 製品安全に関係する製品及び関連する製造工程に携わる要員に対する，組織又は顧客によって特定された教育訓練

j) 製品又は工程の変更は，工程及び製品の変更（ISO 9001の8.3.6参照）による製品安全に関する潜在的影響の評価を含めて，実施前に承認しなければならない．

k) 顧客指定の供給者（8.4.3.1参照）を含む，サプライチェーン全体にわたって製品安全に関する要求事項の連絡

l) サプライチェーン全体にわたって，（最低限）製造ロット単位での製品トレーサビリティ（8.5.2.1参照）
m) 新製品導入に活かす学んだ教訓

注記　特別承認は，安全に関係する内容をもつ

al by the function (typically the customer) that is responsible to approve such documents with safety-related content.

**4.4.2**

See ISO 9001:2015 requirements.

# 5 Leadership

## 5.1 Leadership and commitment

### 5.1.1 General

See ISO 9001:2015 requirements.

#### 5.1.1.1 Corporate responsibility

The organization shall define and implement corporate responsibility policies, including at a minimum an anti-bribery policy, an employee code of conduct, and an ethics escalation policy ("whistle-blowing policy").

#### 5.1.1.2 Process effectiveness and efficiency

Top management shall review the product realization processes and support processes to evaluate and improve their effectiveness and efficiency. The results of the process review activities shall

文書を承認する責任のある機能（通常は顧客）による追加の承認である．

**4.4.2**

ISO 9001:2015 要求事項参照．

**5　リーダーシップ**

**5.1　リーダーシップ及びコミットメント**

**5.1.1　一般**

ISO 9001:2015 要求事項参照．

#### 5.1.1.1　企業責任

組織は，企業責任方針を定め，実施しなければならない．それには，最低限，贈賄防止方針，従業員行動規範及び倫理的上申方針（内部告発方針）を含める．

#### 5.1.1.2　プロセスの有効性及び効率

トップマネジメントは，プロセスの有効性及び効率を評価し改善するために，製品実現プロセス及び支援プロセスをレビューしなければならない．プロセスレビュー活動の結果は，マネジメントレビュー

be included as input to the management review (see Section 9.3.2.1.).

#### 5.1.1.3 Process owners

Top management shall identify process owners who are responsible for managing the organization's processes and related outputs. Process owners shall understand their roles and be competent to perform those roles (see ISO 9001, Section 7.2).

#### 5.1.2 Customer focus

See ISO 9001:2015 requirements.

## 5.2 Policy

### 5.2.1 Establishing the quality policy

See ISO 9001:2015 requirements.

### 5.2.2 Communicating the quality policy

See ISO 9001:2015 requirements.

## 5.3 Organizational roles, responsibilities and authorities

See ISO 9001:2015 requirements.

（9.3.2.1 参照）へのインプットとして含めなければならない．

#### 5.1.1.3　プロセスオーナー

トップマネジメントは，組織のプロセス及び関係するアウトプットをマネジメントする責任をもつプロセスオーナーを特定しなければならない．プロセスオーナーは，自らの役割を理解し，その役割を実行する力量がなければならない（ISO 9001 の 7.2 参照）．

### 5.1.2　顧客重視

ISO 9001:2015 要求事項参照．

## 5.2　方針

### 5.2.1　品質方針の確立

ISO 9001:2015 要求事項参照．

### 5.2.2　品質方針の伝達

ISO 9001:2015 要求事項参照．

## 5.3　組織の役割，責任及び権限

ISO 9001:2015 要求事項参照．

### 5.3.1 Organizational roles, responsibilities, and authorities — supplemental

Top management shall assign personnel with the responsibility and authority to ensure that customer requirements are met. These assignments shall be documented. This includes but is not limited to: the selection of special characteristics, setting quality objectives and related training, corrective and preventive actions, product design and development, capacity analysis, logistics information, customer scorecards, and customer portals.

### 5.3.2 Responsibility and authority for product requirements and corrective actions

Top management shall ensure that:

a) personnel responsible for conformity to product requirements have the authority to stop shipment and stop production to correct quality problems;

   NOTE Due to the process design in some industries, it might not always be possible to stop production immediately. In this case, the affected batch must be contained and

### 5.3.1　組織の役割，責任及び権限—補足

トップマネジメントは，顧客要求事項が満たされることを確実にするために，責任及び権限をもつ要員を任命しなければならない．これらの任命は，文書化しなければならない．これには，特殊特性の選定，品質目標の設定及び関連する教育訓練，是正処置及び予防処置，製品の設計・開発，生産能力分析，物流情報，顧客スコアカード及び顧客ポータルを含めるが，それに限定されない．

### 5.3.2　製品要求事項及び是正処置に対する責任及び権限

トップマネジメントは，次の事項を確実にしなければならない．

a)　製品要求事項への適合に責任を負う要員は，品質問題を是正するために，出荷を停止し，生産を停止する権限をもつ．

注記　産業によっては工程設計のゆえに，必ずしも直ちに生産を停止できない場合がある．この場合，影響のあるバッチは封じ込めて，顧客への出荷

shipment to the customer prevented.

b) personnel with authority and responsibility for corrective action are promptly informed of products or processes that do not conform to requirements to ensure that nonconforming product is not shipped to the customer and that all potential nonconforming product is identified and contained;

c) production operations across all shifts are staffed with personnel in charge of, or delegated responsibility for, ensuring conformity to product requirements.

# 6 Planning

## 6.1 Actions to address risks and opportunities

### 6.1.1 and 6.1.2

See ISO 9001:2015 requirements.

#### 6.1.2.1 Risk analysis

The organization shall include in its risk analysis, at a minimum, lessons learned from product recalls, product audits, field returns and repairs, complaints, scrap, and rework.

を防止しなければならない．

b)　不適合製品が顧客に出荷されないように，また，全ての潜在的不適合製品を識別し封じ込めるために，是正処置に対する権限及び責任をもつ要員に，要求事項に適合しない製品又はプロセスの情報が速やかに報告される．

c)　全てのシフトにわたる生産活動に，製品要求事項への適合を確実にする責任を負う，又はその責任を委任された要員を配置する．

**6　計画**

**6.1　リスク及び機会への取組み**

**6.1.1 及び 6.1.2**

ISO 9001:2015 要求事項参照．

**6.1.2.1　リスク分析**

組織は，最低限，製品のリコールから学んだ教訓，製品監査，市場で起きた回収・修理，苦情，スクラップ及び手直しを，リスク分析に含めなければならない．

The organization shall retain documented information as evidence of the results of risk analysis.

### 6.1.2.2 Preventive action

The organization shall determine and implement action(s) to eliminate the causes of potential nonconformities in order to prevent their occurrence. Preventive actions shall be appropriate to the severity of the potential issues.

The organization shall establish a process to lessen the impact of negative effects of risk including the following:

a) determining potential nonconformities and their causes;
b) evaluating the need for action to prevent occurrence of nonconformities;
c) determining and implementing action needed;
d) documented information of action taken;
e) reviewing the effectiveness of the preventive action taken;
f) utilizing lessons learned to prevent recurrence in similar processes (see ISO 9001, Section 7.1.6).

組織は，リスク分析の結果の証拠として，文書化した情報を保持しなければならない．

### 6.1.2.2　予防処置

組織は，起こり得る不適合が発生することを防止するために，その原因を除去する処置を決め，実施しなければならない．予防処置は，起こり得る問題の重大性に応じたものでなければならない．

組織は，次の事項を含む，リスクの悪影響を及ぼす度合いを減少させるプロセスを確立しなければならない．

a)　起こり得る不適合及びその原因の特定

b)　不適合の発生を予防するための処置の必要性の評価

c)　必要な処置の決定及び実施

d)　とった処置の文書化した情報

e)　とった予防処置の有効性のレビュー

f)　類似プロセスでの再発を防止するために学んだ教訓の活用（ISO 9001 の 7.1.6 参照）

### 6.1.2.3 Contingency plans

The organization shall:

a) identify and evaluate internal and external risks to all manufacturing processes and infrastructure equipment essential to maintain production output and to ensure that customer requirements are met;

b) define contingency plans according to risk and impact to the customer;

c) prepare contingency plans for continuity of supply in the event of any of the following: key equipment failures (also see Section 8.5.6.1.1); interruption from externally provided products, processes, and services; recurring natural disasters; fire; utility interruptions; labour shortages; or infrastructure disruptions;

d) include, as a supplement to the contingency plans, a notification process to the customer and other interested parties for the extent and duration of any situation impacting customer operations;

e) periodically test the contingency plans for effectiveness (e.g., simulations, as appropriate);

### 6.1.2.3　緊急事態対応計画

組織は，次の事項を実施しなければならない．

a)　顧客要求事項が満たされることを確実にし，生産からのアウトプットを維持するのに不可欠な全ての製造工程及びインフラストラクチャの設備に対する内部及び外部のリスクを特定し評価する．

b)　顧客へのリスク及び影響に従って，緊急事態対応計画を定める．

c)　次の事態において，供給の継続のために緊急事態対応計画を準備する．主要設備の故障（8.5.6.1.1 も参照），外部から提供される製品，プロセス，及びサービスの中断，繰り返し発生する自然災害，火事，ユーティリティの停止，労働力不足，又はインフラストラクチャ障害

d)　顧客の操業に影響するいかなる状況も，その程度及び期間に対して，顧客及び他の利害関係者への通知プロセスを，緊急事態対応計画の補完として含める．

e)　有効性（例　必要に応じて，シミュレーション）について，定期的に緊急事態対応計画をテストする．

f) conduct contingency plan reviews (at a minimum annually) using a multidisciplinary team including top management, and update as required;

g) document the contingency plans and retain documented information describing any revision(s), including the person(s) who authorized the change(s).

The contingency plans shall include provisions to validate that the manufactured product continues to meet customer specifications after the re-start of production following an emergency in which production was stopped and if the regular shutdown processes were not followed.

## 6.2 Quality objectives and planning to achieve them

### 6.2.1 and 6.2.2

See ISO 9001:2015 requirements.

#### 6.2.2.1 Quality objectives and planning to achieve them — supplemental

Top management shall ensure that quality objec-

f)　トップマネジメントを含む部門横断チームによって，緊急事態対応計画のレビューを行い（最低限，年次で），必要に応じて更新する．

g)　緊急事態対応計画を文書化し，また，変更を正式許可した人を含む，いかなる改訂をも記述した文書化した情報を保持する．

緊急事態対応計画には，生産が停止した緊急事態の後で生産を再稼働したとき及び正規のシャットダウンプロセスがとられなかった場合，製造された製品が引き続き顧客仕様を満たすことの妥当性確認条項を含めなければならない．

## 6.2　品質目標及びそれを達成するための計画策定

### 6.2.1 及び 6.2.2

ISO 9001:2015 要求事項参照．

#### 6.2.2.1　品質目標及びそれを達成するための計画策定—補足

トップマネジメントは，組織全体にわたって，関

tives to meet customer requirements are defined, established, and maintained for relevant functions, processes, and levels throughout the organization.

The results of the organization's review regarding interested parties and their relevant requirements shall be considered when the organization establishes its annual (at a minimum) quality objectives and related performance targets (internal and external).

### 6.3 Planning of changes

See ISO 9001:2015 requirements.

## 7 Support

### 7.1 Resources

#### 7.1.1 General

See ISO 9001:2015 requirements.

#### 7.1.2 People

See ISO 9001:2015 requirements.

#### 7.1.3 Infrastructure

See ISO 9001:2015 requirements.

連する機能，プロセス及び階層において，顧客要求事項を満たす品質目標を定め，確立し及び維持することを確実にしなければならない．

利害関係者及びその関連する要求事項に関する組織のレビューの結果は，組織が最低限，年次の品質目標及び関係するパフォーマンス目標（内部及び外部）を確立する際に，考慮しなければならない．

**6.3　変更の計画**

ISO 9001:2015 要求事項参照．

**7　支援**

**7.1　資源**

**7.1.1　一般**

ISO 9001:2015 要求事項参照．

**7.1.2　人々**

ISO 9001:2015 要求事項参照．

**7.1.3　インフラストラクチャ**

ISO 9001:2015 要求事項参照．

### 7.1.3.1 Plant, facility, and equipment planning

The organization shall use a multidisciplinary approach including risk identification and risk mitigation methods for developing and improving plant, facility, and equipment plans. In designing plant layouts, the organization shall:

a) optimize material flow, material handling, and value-added use of floor space including control of nonconforming product, and
b) facilitate synchronous material flow, as applicable.

Methods shall be developed and implemented to evaluate manufacturing feasibility for new product or new operations. Manufacturing feasibility assessments shall include capacity planning. These methods shall also be applicable for evaluating proposed changes to existing operations.

The organization shall maintain process effectiveness, including periodic re-evaluation relative to risk, to incorporate any changes made during process approval, control plan maintenance (see Sec-

#### 7.1.3.1　工場，施設及び設備の計画

組織は，工場，施設及び設備の計画を策定し改善するために，リスク特定及びリスク緩和の方法を含めて，部門横断的アプローチを用いなければならない．工場レイアウトを設計する際は，組織は次の事項を実施しなければならない．

a)　不適合製品の管理を含む，材料の流れ，材料の取扱い及び現場スペースの付加価値のある活用を最適化する．

b)　該当する場合には，必ず，同期のとれた材料の流れを促進する．

新製品及び新運用に対する製造フィージビリティを評価するために，方法を開発し，実施しなければならない．製造フィージビリティ評価には，生産能力計画を含めなければならない．これらの方法は，既存の運用への提案された変更を評価することにも適用可能でなければならない．

組織は，リスクに関連する定期的再評価を含めて，工程承認中になされた変更，コントロールプランの維持（8.5.1.1 参照）及び作業の段取り替え検証（8.5.1.3 参照）を取り入れるために，工程の有

tion 8.5.1.1), and verification of job set-ups (see Section 8.5.1.3).

Assessments of manufacturing feasibility and evaluation of capacity planning shall be inputs to management reviews (see ISO 9001, Section 9.3).

NOTE 1 These requirements should include the application of lean manufacturing principles.
NOTE 2 These requirements should apply to on-site supplier activities, as applicable.

### 7.1.4 Environment for the operation of processes

See ISO 9001:2015 requirements.

NOTE Where third-party certification to ISO 45001 (or equivalent) is recognized, it may be used to demonstrate the organization's conformity to the personnel safety aspects of this requirement.

#### 7.1.4.1 Environment for the operation of processes — supplemental

効性を維持しなければならない.

製造フィージビリティ評価及び生産能力評価は,マネジメントレビューへのインプットとしなければならない(ISO 9001 の 9.3 参照).

注記 1　これらの要求事項には,リーン生産の原則の適用を含めることが望ましい.

注記 2　これらの要求事項は,該当する場合には,必ず,サイト内供給者の活動に適用することが望ましい.

**7.1.4　プロセスの運用に関する環境**

ISO 9001:2015 要求事項参照.

注記　ISO 45001(又はそれに相当するもの)への第三者認証が承認されれば,この要求事項の要員安全の側面に対する組織の適合を実証するために用いてもよい.

**7.1.4.1　プロセスの運用に関する環境—補足**

The organization shall maintain its premises in a state of order, cleanliness, and repair that is consistent with the product and manufacturing process needs.

### 7.1.5 Monitoring and measuring resources

#### 7.1.5.1 General

See ISO 9001:2015 requirements.

##### 7.1.5.1.1 Measurement system analysis

Statistical studies shall be conducted to analyse the variation present in the results of each type of inspection, measurement, and test equipment system identified in the control plan. The analytical methods and acceptance criteria used shall conform to those in reference manuals on measurement systems analysis. Other analytical methods and acceptance criteria may be used if approved by the customer.

Records of customer acceptance of alternative methods shall be retained along with results from alternative measurement systems analysis (see

組織は，製品及び製造工程のニーズに合わせて，事業所を整頓され清潔で手入れされた状態に維持しなければならない．

**7.1.5　監視及び測定のための資源**

**7.1.5.1　一般**

ISO 9001:2015 要求事項参照．

**7.1.5.1.1　測定システム解析**

コントロールプランに特定されている各種の検査，測定及び試験設備システムの結果に存在するばらつきを解析するために，統計的調査を実施しなければならない．使用する解析方法及び合否判定基準は，測定システム解析に関するレファレンスマニュアルに適合しなければならない．顧客が承認した場合は，他の解析方法及び合否判定基準を使用してもよい．

代替方法に対する顧客承諾の記録は，代替の測定システム解析の結果とともに保持しなければならない（9.1.1.1 参照）．

Section 9.1.1.1).

NOTE Prioritization of MSA studies should focus on critical or special product or process characteristics.

#### 7.1.5.2 Measurement traceability

See ISO 9001:2015 requirements.

NOTE A number or another identifier traceable to the device calibration record meets the intent of the requirements in ISO 9001:2015.

#### 7.1.5.2.1 Calibration / verification records

The organization shall have a documented process for managing calibration/verification records. Records of the calibration/verification activity for all gauges and measuring and test equipment (including employee-owned equipment relevant for measuring, customer-owned equipment, or on-site supplier-owned equipment) needed to provide evidence of conformity to internal requirements, legislative and regulatory requirements, and customer-defined requirements shall be retained.

注記　MSA 調査の優先順位は，製品若しくは工程の重大特性又は特殊特性を重視することが望ましい．

#### 7.1.5.2　測定のトレーサビリティ

ISO 9001:2015 要求事項参照．

注記　機器の校正記録に対してトレーサブルな番号又は他の識別子は，ISO 9001:2015 における要求事項の意図を満たす．

#### 7.1.5.2.1　校正／検証の記録

組織は，校正／検証の記録をマネジメントする文書化したプロセスをもたなければならない．内部要求事項，法令・規制要求事項及び顧客が定めた要求事項への適合の証拠を提供するために必要な，全てのゲージ，測定機器及び試験設備（測定に関連する従業員所有の機器，顧客所有の機器，サイト内供給者所有の機器を含む．）に対する校正／検証の活動の記録は，保持しなければならない．

The organization shall ensure that calibration/verification activities and records shall include the following details:

a) revisions following engineering changes that impact measurement systems;

b) any out-of-specification readings as received for calibration/verification;

c) an assessment of the risk of the intended use of the product caused by the out-of-specification condition;

d) when a piece of inspection measurement and test equipment is found to be out of calibration or defective during its planned verification or calibration or during its use, documented information on the validity of previous measurement results obtained with this piece of inspection measurement and test equipment shall be retained, including the associated standard's last calibration date and the next due date on the calibration report;

e) notification to the customer if suspect product or material has been shipped;

f) statements of conformity to specification after calibration/verification;

組織は，校正／検証の活動及び記録には次の詳細事項を含めなければならないことを確実にしなければならない．

a)　測定システムに影響する，設計変更による改訂

b)　校正／検証のために受け入れた状態で，仕様外れの値

c)　仕様外れ状態によって引き起こされ得る製品の意図した用途に対するリスクの評価

d)　検査測定及び試験設備が，計画した検証中若しくは校正中に，又はその使用中に，校正外れ又は故障が発見された場合，この検査測定及び試験設備によって得られた以前の測定結果の妥当性に関する文書化した情報を，校正報告書に関連する標準器の最後の校正を行った日付及び次の校正が必要になる期限を含めて，保持しなければならない．

e)　疑わしい製品又は材料が出荷された場合の顧客への通知

f)　校正／検証後の，仕様への適合表明

g) verification that the software version used for product and process control is as specified;

h) records of the calibration and maintenance activities for all gauging (including employee-owned equipment, customer-owned equipment, or on-site supplier-owned equipment);

i) production-related software verification used for product and process control (including software installed on employee-owned equipment, customer-owned equipment, or on-site supplier-owned equipment).

### 7.1.5.3 Laboratory requirements

#### 7.1.5.3.1 Internal laboratory

An organization's internal laboratory facility shall have a defined scope that includes its capability to perform the required inspection, test, or calibration services. This laboratory scope shall be included in the quality management system documentation. The laboratory shall specify and implement, as a minimum, requirements for:

a) adequacy of the laboratory technical procedures;

b) competency of the laboratory personnel;

g) 製品及び工程の管理に使用されるソフトウェアのバージョンが指示どおりであることの検証
h) 全てのゲージ（従業員所有の機器，顧客所有の機器，サイト内供給者所有の機器を含む）に対する校正及び保全活動の記録

i) 製品及び工程の管理に使用される（従業員所有の機器，顧客所有の機器，サイト内供給者所有の機器にインストールされたソフトウェアを含む）生産に関係するソフトウェアの検証

**7.1.5.3 試験所要求事項**

**7.1.5.3.1 内部試験所**

組織内部の試験所施設は，要求される検査，試験又は校正サービスを実行する能力を含む，定められた適用範囲をもたなければならない．この試験所適用範囲は，品質マネジメントシステム文書に含めなければならない．試験所は，最低限，次の事項に対する要求事項を規定し，実施しなければならない．

a) 試験所の技術手順の適切性

b) 試験所要員の力量

c) testing of the product;
d) capability to perform these services correctly, traceable to the relevant process standard (such as ASTM, EN, etc.); when no national or international standard(s) is available, the organization shall define and implement a methodology to verify measurement system capability;
e) customer requirements, if any;
f) review of the related records.

NOTE Third-party accreditation to ISO/IEC 17025 (or equivalent) may be used to demonstrate the organization's in-house laboratory conformity to this requirement.

### 7.1.5.3.2 External laboratory

External/commercial/independent laboratory facilities used for inspection, test, or calibration services by the organization shall have a defined laboratory scope that includes the capability to perform the required inspection, test, or calibration, and either:

— the laboratory shall be accredited to ISO/IEC

c)　製品の試験

d)　該当するプロセス規格（ASTM，EN などのような）にトレーサブルな形で，これらのサービスを正確に実行する能力．国家標準又は国際標準が存在しない場合，組織は，測定システムの能力を検証する手法を定めて実施しなければならない．

e)　もしあれば，顧客要求事項

f)　関係する記録のレビュー

注記　ISO/IEC 17025（又はそれに相当するもの）に対する第三者認定を，組織の内部試験所がこの要求事項に適合していることの実証に使用してもよい．

**7.1.5.3.2　外部試験所**

組織が検査，試験，又は校正サービスに使用する，外部／商用／独立の試験所施設は，要求される検査，試験，又は校正を実行する能力を含む，定められた試験所適用範囲をもたなければならない．また，次の事項のいずれかを満たさなければならない．

—　試験所は，ISO/IEC 17025 又はこれに相当す

17025 or national equivalent and include the relevant inspection, test, or calibration service in the scope of the accreditation (certificate); the certificate of calibration or test report shall include the mark of a national accreditation body; or

— there shall be evidence that the external laboratory is acceptable to the customer.

NOTE Such evidence may be demonstrated by customer assessment, for example, or by customer-approved second-party assessment that the laboratory meets the intent of ISO/IEC 17025 or national equivalent. The second-party assessment may be performed by the organization assessing the laboratory using a customer-approved method of assessment.

Calibration services may be performed by the equipment manufacturer when a qualified laboratory is not available for a given piece of equipment. In such cases, the organization shall ensure that the requirements listed in Section 7.1.5.3.1 have been met.

る国内基準に認定され，該当する検査，試験，又は校正サービスを認定（認証書）の適用範囲に含めなければならない．校正又は試験報告書の認証書は，国家認定機関のマークを含んでいなければならない．

— 外部試験所が顧客に受け入れられることができるとの証拠がなければならない．

注記　そのような証拠は，例えば，試験所がISO/IEC 17025 又はこれに相当する国内基準の意図を満たすとの顧客評価，又は顧客が認めた第二者評価によって，実証してもよい．顧客が認めた評価方法を使用して試験所を評価する組織によって第二者評価を，実行してもよい．

ある機器に対して認定された試験所を利用できない場合，校正サービスは，機器の製造業者によって実行してもよい．この場合，組織は，7.1.5.3.1 に記載された要求事項を満たすことを確実にしなければならない．

Use of calibration services, other than by qualified (or customer accepted) laboratories, may be subject to government regulatory confirmation, if required.

### 7.1.6 Organizational knowledge

See ISO 9001:2015 requirements.

## 7.2 Competence

See ISO 9001:2015 requirements.

### 7.2.1 Competence — supplemental

The organization shall establish and maintain a documented process(es) for identifying training needs including awareness (see Section 7.3.1) and achieving competence of all personnel performing activities affecting conformity to product and process requirements. Personnel performing specific assigned tasks shall be qualified, as required, with particular attention to the satisfaction of customer requirements.

### 7.2.2 Competence — on-the-job training

The organization shall provide on-the-job train-

校正サービスの使用は，認定された（又は顧客が認めた）試験所以外による場合，要求されれば，政府規制の確認の対象となる場合がある．

**7.1.6 組織の知識**

ISO 9001:2015 要求事項参照．

**7.2 力量**

ISO 9001:2015 要求事項参照．

**7.2.1 力量—補足**

組織は，製品及びプロセス要求事項への適合に影響する活動に従事する全ての要員の，認識（7.3.1 参照）を含めて，教育訓練のニーズ及び達成すべき力量を明確にする文書化したプロセスを確立し，維持しなければならない．必要に応じて，顧客要求事項を満たすことに特に配慮して，特定の業務に従事する要員を適格性確認しなければならない．

**7.2.2 力量—業務を通じた教育訓練（OJT）**

組織は，品質要求事項への適合，内部要求事項，

ing (which shall include customer requirements training) for personnel in any new or modified responsibilities affecting conformity to quality requirements, internal requirements, regulatory or legislative requirements; this shall include contract or agency personnel. The level of detail required for on-the-job training shall be commensurate with the level of education the personnel possess and the complexity of the task(s) they are required to perform for their daily work. Persons whose work can affect quality shall be informed about the consequences of nonconformity to customer requirements.

### 7.2.3 Internal auditor competency

The organization shall have a documented process(es) to verify that internal auditors are competent, taking into account any customer-specific requirements. For additional guidance on auditor competencies, refer to ISO 19011. The organization shall maintain a list of qualified internal auditors.

Quality management system auditors, manu-

規制又は法令要求事項に影響する，新規の又は変更された責任を負う要員に対し，業務を通じた教育訓練（OJT）（これには顧客要求事項の教育訓練を含めなければならない.）を提供しなければならない．これには，契約又は派遣の要員を含めなければならない．業務を通じた教育訓練（OJT）に対する詳細な要求レベルは，要員が有する教育及び日常業務を実行するために必要な任務の複雑さのレベルに見合っていなければならない．品質に影響し得る仕事に従事する要員には，顧客要求事項に対する不適合の因果関係について，知らせなければならない．

### 7.2.3　内部監査員の力量

組織は，顧客固有要求事項を考慮に入れて，内部監査員が力量をもつことを検証する文書化したプロセスをもたなければならない．監査員の力量に関する追加の手引には，ISO 19011 を参照する．組織は，資格をもつ内部監査員のリストを維持しなければならない．

品質マネジメントシステム監査員，製造工程監査

facturing process auditors, and product auditors shall all be able to demonstrate the following minimum competencies:

a) understanding of the automotive process approach for auditing, including risk-based thinking;
b) understanding of applicable customer-specific requirements;
c) understanding of applicable ISO 9001 and IATF 16949 requirements related to the scope of the audit;
d) understanding of applicable core tool requirements related to the scope of the audit;
e) understanding how to plan, conduct, report, and close out audit findings.

Additionally, manufacturing process auditors shall demonstrate technical understanding of the relevant manufacturing process(es) to be audited, including process risk analysis (such as PFMEA) and control plan. Product auditors shall demonstrate competence in understanding product requirements and use of relevant measuring and test equipment to verify product conformity.

員及び製品監査員は，全て，次の最低限の力量を実証できなければならない．

a)　リスクに基づく考え方を含む，監査に対する自動車産業プロセスアプローチの理解

b)　該当する顧客固有要求事項の理解

c)　監査範囲に関係する，該当する ISO 9001 及び IATF 16949 要求事項の理解

d)　監査範囲に関係する，該当するコアツール要求事項の理解

e)　計画，実施，報告及び監査所見の完了の仕方の理解

さらに，製造工程監査員は，監査対象となる該当する製造工程の，工程リスク分析（PFMEA のような）及びコントロールプランを含む，専門的理解を実証しなければならない．製品監査員は，製品の適合性を検証するために，製品要求事項の理解，並びに該当する測定及び試験設備の使用において，力量を実証しなければならない．

Where training is provided to achieve competency, documented information shall be retained to demonstrate the trainer's competency with the above requirements.

Maintenance of and improvement in internal auditor competence shall be demonstrated through:

f) executing a minimum number of audits per year, as defined by the organization; and
g) maintaining knowledge of relevant requirements based on internal changes (e.g., process technology, product technology) and external changes (e.g., ISO 9001, IATF 16949, core tools, and customer specific requirements).

### 7.2.4 Second-party auditor competency

The organization shall demonstrate the competence of the auditors undertaking the second-party audits. Second-party auditors shall meet customer specific requirements for auditor qualification and demonstrate the minimum following core competencies, including understanding of:

a) the automotive process approach to auditing, including risk based thinking;

力量を獲得するために教育訓練が提供される際は，上記要求事項を備えたトレーナーの力量を実証するために文書化した情報を保持しなければならない．

内部監査員の力量における維持及び改善は，次の事項を通じて実証しなければならない．

f)　組織が定める，年間最低回数の監査の実施

g)　内部変化（例　工程技術，製品技術）及び外部変化（例　ISO 9001，IATF 16949，コアツール及び顧客固有要求事項）に基づく，該当する要求事項の知識の維持

### 7.2.4　第二者監査員の力量

組織は，第二者監査を実施する監査員の力量を実証しなければならない．第二者監査員は，監査員の適格性確認に対する顧客固有要求事項を満たし，次の事項の理解を含む，最低限の核となる次の力量を実証しなければならない．

a)　リスクに基づく考え方を含む，監査に対する自動車産業プロセスアプローチ

b) applicable customer and organization specific requirements;
c) applicable ISO 9001 and IATF 16949 requirements related to the scope of the audit;
d) applicable manufacturing process(es) to be audited, including PFMEA and control plan;
e) applicable core tool requirements related to the scope of the audit;
f) how to plan, conduct, prepare audit reports, and close out audit findings.

## 7.3 Awareness

See ISO 9001:2015 requirements.

### 7.3.1 Awareness — supplemental

The organization shall maintain documented information that demonstrates that all employees are aware of their impact on product quality and the importance of their activities in achieving, maintaining, and improving quality, including customer requirements and the risks involved for the customer with non-conforming product.

b) 該当する顧客及び組織の固有要求事項

c) 監査範囲に関係する，該当する ISO 9001 及び IATF 16949 要求事項

d) PFMEA 及びコントロールプランを含む，監査対象となる製造工程

e) 監査範囲に関係する，該当するコアツール要求事項

f) 計画，実施，監査報告書の準備及び監査所見の完了の仕方

**7.3 認識**

ISO 9001:2015 要求事項参照.

### 7.3.1 認識—補足

組織は，全ての従業員が，顧客要求事項及び不適合製品に関わる顧客のリスクを含む，従業員が製品品質に及ぼす影響，並びに品質を達成し，維持し及び改善するために行う活動の重要性を認識することを実証する，文書化した情報を維持しなければならない.

### 7.3.2 Employee motivation and empowerment

The organization shall maintain a documented process(es) to motivate employees to achieve quality objectives, to make continual improvements, and to create an environment that promotes innovation. The process shall include the promotion of quality and technological awareness throughout the whole organization.

## 7.4 Communication

See ISO 9001:2015 requirements.

## 7.5 Documented information

### 7.5.1 General

See ISO 9001:2015 requirements.

#### 7.5.1.1 Quality management system documentation

The organization's quality management system shall be documented and include a quality manual, which can be a series of documents (electronic or hard copy).

### 7.3.2 従業員の動機付け及びエンパワーメント

組織は，品質目標を達成し，継続的改善を行い及び革新を促進する環境を創造するための，従業員を動機付ける文書化したプロセスを維持しなければならない．そのプロセスには，組織全体にわたって品質及び技術的認識を促進することを含めなければならない．

## 7.4 コミュニケーション

ISO 9001:2015 要求事項参照．

## 7.5 文書化した情報

### 7.5.1 一般

ISO 9001:2015 要求事項参照．

#### 7.5.1.1 品質マネジメントシステムの文書類

組織の品質マネジメントシステムは，文書化し，品質マニュアルに含めなければならない，その品質マニュアルは一連の文書（電子版又は印刷版）であってもよい．

The format and structure of the quality manual is at the discretion of the organization and will depend on the organization's size, culture, and complexity. If a series of documents is used, then a list shall be retained of the documents that comprise the quality manual for the organization.

The quality manual shall include, at a minimum, the following:

a) the scope of the quality management system, including details of and justification for any exclusions;
b) documented processes established for the quality management system, or reference to them;
c) the organization's processes and their sequence and interactions (inputs and outputs), including type and extent of control of any outsourced processes;
d) a document (i.e., matrix) indicating where within the organization's quality management system their customer-specific requirements are addressed.

品質マニュアルの様式及び構成は，組織の裁量により，また，組織の規模，文化及び複雑さによって決まる．一連の文書が使用されるならば，組織の品質マニュアルを構成する文書のリストを保持しなければならない．

品質マニュアルには，最低限，次の事項を含めなければならない．

a)　品質マネジメントシステムの適用範囲．除外がある場合には，除外の詳細，及び除外を正当とする理由

b)　品質マネジメントシステムについて確立された，文書化したプロセス，又はそれらを参照できる情報

c)　アウトソースしたプロセスの管理の方式及び程度を含む，組織のプロセス並びにそれらの順序及び相互作用（インプット及びアウトプット）

d)　組織の品質マネジメントシステムの中の，どこで顧客固有要求事項に取り組んでいるかを示す文書（すなわち，マトリックス）

NOTE A matrix of how the requirements of this Automotive QMS standard are addressed by the organization's processes may be used to assist with linkages of the organization's processes and this Automotive QMS.

**7.5.2 Creating and updating**

See ISO 9001:2015 requirements.

**7.5.3 Control of documented information**

**7.5.3.1 and 7.5.3.2**

See ISO 9001:2015 requirements.

### 7.5.3.2.1 Record retention

The organization shall define, document, and implement a record retention policy. The control of records shall satisfy statutory, regulatory, organizational, and customer requirements.

Production part approvals, tooling records (including maintenance and ownership), product and process design records, purchase orders (if applicable), or contracts and amendments shall be re-

注記 この自動車産業QMS規格の要求事項が，どのように組織のプロセスによって取り組まれているかを示すマトリックスは，組織のプロセスとこの自動車産業QMSとのつながりを支援するために利用してもよい．

**7.5.2 作成及び更新**

ISO 9001:2015要求事項参照．

**7.5.3 文書化した情報の管理**

**7.5.3.1 及び 7.5.3.2**

ISO 9001:2015要求事項参照．

**7.5.3.2.1 記録の保管**

組織は，記録保管方針を定め，文書化し，実施しなければならない．記録の管理は，法令，規制，組織及び顧客要求事項を満たさなければならない．

生産部品承認，治工具の記録（保全及び保有者を含む），製品設計及び工程設計の記録，購買注文書（該当する場合には，必ず），契約書及びその修正事項は，製品が生産及びサービス要求事項に対して有

tained for the length of time that the product is active for production and service requirements, plus one calendar year, unless otherwise specified by the customer or regulatory agency.

NOTE Production part approval documented information may include approved product, applicable test equipment records, or approved test data.

### 7.5.3.2.2 Engineering specifications

The organization shall have a documented process describing the review, distribution, and implementation of all customer engineering standards/specifications and related revisions based on customer schedules, as required.

When an engineering standard/specification change results in a product design change, refer to the requirements in ISO 9001, Section 8.3.6. When an engineering standard/specification change results in a product realization process change, refer to the requirements in Section 8.5.6.1. The organization shall retain a record of

効である期間に加えて1暦年，保持しなければならない．ただし，顧客又は規制当局によって規定されたときは，この限りでない．

注記 生産部品承認の文書化した情報には，承認された製品，該当する設備の記録，又は承認された試験データを含めてもよい．

**7.5.3.2.2 技術仕様書**

組織は，顧客の全ての技術規格／仕様書及び関係する改訂に対して，要求される顧客スケジュールに基づいて，レビュー，配付及び実施を記述した文書化したプロセスをもたなければならない．

技術規格／仕様書の変更が，製品設計変更になる場合は，ISO 9001 の 8.3.6 の要求事項を参照する．技術規格／仕様書の変更が，製品実現プロセスの変更になる場合は，8.5.6.1 の要求事項を参照する．組織は，生産において実施された各変更の日付の記録を保持しなければならない．実施には，更新された文書を含めなければならない．

the date on which each change is implemented in production. Implementation shall include updated documents.

Review should be completed within 10 working days of receipt of notification of engineering standards/specifications changes.

NOTE A change in these standards/specifications may require an updated record of customer production part approval when these specifications are referenced on the design record or if they affect documents of the production part approval process, such as control plan, risk analysis (such as FMEAs), etc.

# 8 Operation

## 8.1 Operational planning and control

See ISO 9001:2015 requirements.

### 8.1.1 Operational planning and control — supplemental

When planning for product realization, the following topics shall be included:

レビューは，技術規格／仕様書の変更を受領してから，10 稼働日内に完了することが望ましい.

注記　技術規格／仕様書の変更は，仕様書が設計記録に引用されている，又は，コントロールプラン，リスク分析（FMEA のような），などのような生産部品承認プロセス文書に影響する場合，顧客の生産部品承認の更新された記録が要求される場合がある.

**8　運用**

**8.1　運用の計画及び管理**

ISO 9001:2015 要求事項参照.

### 8.1.1　運用の計画及び管理—補足

製品実現の計画をする際は，次の事項を含めなければならない.

a) customer product requirements and technical specifications;
b) logistics requirements;
c) manufacturing feasibility;
d) project planning (refer to ISO 9001, Section 8.3.2);
e) acceptance criteria.

The resources identified in ISO 9001, Section 8.1 c), refer to the required verification, validation, monitoring, measurement, inspection, and test activities specific to the product and the criteria for product acceptance.

### 8.1.2 Confidentiality

The organization shall ensure the confidentiality of customer-contracted products and projects under development, including related product information.

## 8.2 Requirements for products and services

### 8.2.1 Customer communication

See ISO 9001:2015 requirements.

a)　顧客の製品要求事項及び技術仕様書

b)　物流要求事項

c)　製造フィージビリティ

d)　プロジェクト計画（ISO 9001 の 8.3.2 参照）

e)　合否判定基準

ISO 9001 の 8.1 c)に特定される資源は，製品及び製品の合否判定基準に固有の，要求される検証，妥当性確認，監視，測定，検査及び試験活動について述べている．

### 8.1.2　機密保持

組織は，顧客と契約した開発中の製品及びプロジェクト，並びに関係製品情報の機密保持を確実にしなければならない．

## 8.2　製品及びサービスに関する要求事項

### 8.2.1　顧客とのコミュニケーション

ISO 9001:2015 要求事項参照．

### 8.2.1.1 Customer communication — supplemental

Written or verbal communication shall be in the language agreed with the customer. The organization shall have the ability to communicate necessary information, including data in a customer-specified computer language and format (e.g., computer-aided design data, electronic data interchange).

### 8.2.2 Determining the requirements for products and services

See ISO 9001:2015 requirements.

### 8.2.2.1 Determining the requirements for products and services — supplemental

These requirements shall include recycling, environmental impact, and characteristics identified as a result of the organization's knowledge of the product and manufacturing processes.

Compliance to ISO 9001, Section 8.2.2 item a) 1), shall include but not be limited to the following: all applicable government, safety, and environ-

### 8.2.1.1　顧客とのコミュニケーション—補足

記述された又は口頭のコミュニケーションは，顧客と合意した言語によらなければならない．組織は，顧客に規定されたコンピュータ言語及び書式（例　CAD データ，電子データ交換）を含めて，必要な情報を伝達する能力をもたなければならない．

### 8.2.2　製品及びサービスに関する要求事項の明確化

ISO 9001:2015 要求事項参照.

### 8.2.2.1　製品及びサービスに関する要求事項の明確化—補足

これらの要求事項には，製品及び製造工程について組織の知識の結果としてリサイクル，環境影響及び特性を含めなければならない．

ISO 9001 の 8.2.2 a)1)への適合には，材料の入手，保管，取扱い，リサイクル，除去，又は廃棄に関係する，全ての該当する政府の，安全規制及び環

mental regulations related to acquisition, storage, handling, recycling, elimination, or disposal of material.

**8.2.3 Review of the requirements for products and services**

**8.2.3.1**

See ISO 9001:2015 requirements.

**8.2.3.1.1 Review of the requirements for products and services — supplemental**

The organization shall retain documented evidence of a customer-authorized waiver for the requirements stated in ISO 9001, Section 8.2.3.1, for a formal review.

**8.2.3.1.2 Customer-designated special characteristics**

The organization shall conform to customer requirements for designation, approval documentation, and control of special characteristics.

**8.2.3.1.3 Organization manufacturing feasibility**

境規制を含めなければならない．しかし，それに限定されない．

### 8.2.3　製品及びサービスに関する要求事項のレビュー

#### 8.2.3.1

ISO 9001:2015 要求事項参照．

##### 8.2.3.1.1　製品及びサービスに関する要求事項のレビュー—補足

組織は，正式なレビューのための，ISO 9001 の 8.2.3.1 に述べられている要求事項に対する顧客が正式許可した免除の，文書化した証拠を保持しなければならない．

##### 8.2.3.1.2　顧客指定の特殊特性

組織は，特殊特性の指定，承認文書及び管理に対する顧客要求事項に適合しなければならない．

##### 8.2.3.1.3　組織の製造フィージビリティ

The organization shall utilize a multidisciplinary approach to conduct an analysis to determine if it is feasible that the organization's manufacturing processes are capable of consistently producing product that meets all of the engineering and capacity requirements specified by the customer. The organization shall conduct this feasibility analysis for any manufacturing or product technology new to the organization and for any changed manufacturing process or product design.

Additionally, the organization should validate through production runs, benchmarking studies, or other appropriate methods, their ability to make product to specifications at the required rate.

**8.2.3.2**

See ISO 9001:2015 requirements.

**8.2.4 Changes to requirements for products and services**

See ISO 9001:2015 requirements.

組織は，組織の製造工程が一貫して，顧客の規定した全ての技術及び生産能力の要求事項を満たす製品を生産できることが実現可能か否かを判定するための分析を実施するために，部門横断的アプローチを利用しなければならない．組織は，このフィージビリティ分析を，組織にとって新規の製造技術又は製品技術に対して及び変更された製造工程又は製品設計に対して実施しなければならない．

加えて，組織は，生産稼働，ベンチマーキング調査，又は他の適切な方法で，仕様どおりの製品を要求される速度で生産する能力の妥当性確認を行うことが望ましい．

**8.2.3.2**

ISO 9001:2015 要求事項参照．

**8.2.4　製品及びサービスに関する要求事項の変更**

ISO 9001:2015 要求事項参照．

## 8.3 Design and development of products and services

### 8.3.1 General

See ISO 9001:2015 requirements.

#### 8.3.1.1 Design and development of products and services — supplemental

The requirements of ISO 9001, Section 8.3.1, shall apply to product and manufacturing process design and development and shall focus on error prevention rather than detection.

The organization shall document the design and development process.

### 8.3.2 Design and development planning

See ISO 9001:2015 requirements.

#### 8.3.2.1 Design and development planning — supplemental

The organization shall ensure that design and development planning includes all affected stakeholders within the organization and, as appropriate, its supply chain. Examples of areas for using

## 8.3 製品及びサービスの設計・開発

### 8.3.1 一般

ISO 9001:2015 要求事項参照.

#### 8.3.1.1 製品及びサービスの設計・開発―補足

ISO 9001 の 8.3.1 の要求事項は，製品及び製造工程の設計・開発に適用し，不具合の検出よりも不具合の予防を重視しなければならない.

組織は，設計・開発プロセスを文書化しなければならない.

### 8.3.2 設計・開発の計画

ISO 9001:2015 要求事項参照.

#### 8.3.2.1 設計・開発の計画―補足

組織は，設計・開発プロセスに，影響を受ける全ての組織内の利害関係者及び，必要に応じて，サプライチェーンを含めることを確実にしなければならない. そのような部門横断的アプローチを用いる領

such a multidisciplinary approach include but are not limited to the following:

a) project management (for example, APQP or VDA-RGA);
b) product and manufacturing process design activities (for example, DFM and DFA), such as consideration of the use of alternative designs and manufacturing processes;
c) development and review of product design risk analysis (FMEAs), including actions to reduce potential risks;
d) development and review of manufacturing process risk analysis (for example, FMEAs, process flows, control plans, and standard work instructions).

NOTE A multidisciplinary approach typically includes the organization's design, manufacturing, engineering, quality, production, purchasing, supplier, maintenance, and other appropriate functions.

### 8.3.2.2 Product design skills

The organization shall ensure that personnel with

域の例には，次の事項がある．しかし，それに限定されない．

a)　プロジェクトマネジメント（例えば，APQP 又は VDA-RGA）

b)　代替の設計提案及び製造工程案の使用を検討するような，製品設計及び製造工程設計の活動（例えば，DFM 及び DFA）

c)　潜在的リスクを低減する処置を含む，製品設計リスク分析（FMEA）の実施及びレビュー

d)　製造工程リスク分析の実施及びレビュー（例えば，FMEA，工程フロー，コントロールプラン及び標準作業指示書）

注記　部門横断的アプローチには，通常，組織の設計，製造，技術，品質，生産，購買，供給者，保全及び他の適切な部門を含める．

### 8.3.2.2　製品設計の技能

組織は，製品設計の責任を負う要員が，設計要求

product design responsibility are competent to achieve design requirements and are skilled in applicable product design tools and techniques. Applicable tools and techniques shall be identified by the organization.

NOTE An example of product design skills is the application of digitized mathematically based data.

**8.3.2.3 Development of products with embedded software**

The organization shall use a process for quality assurance for their products with internally developed embedded software. A software development assessment methodology shall be utilized to assess the organization's software development process. Using prioritization based on risk and potential impact to the customer, the organization shall retain documented information of a software development capability self-assessment.

The organization shall include software development within the scope of their internal audit pro-

事項を実現する力量をもち，適用されるツール及び手法の技能をもつことを確実にしなければならない．組織は，適用されるツール及び手法を明確にしなければならない．

注記　製品設計の技能の例の一つとして，デジタル化された数学的なデータの適用がある．

**8.3.2.3　組込みソフトウェアをもつ製品の開発**

組織は，内部で開発された組込みソフトウェアをもつ製品に対する品質保証のプロセスを用いなければならない．ソフトウェア開発評価の方法論を，組織のソフトウェア開発プロセスを評価するために利用しなければならない．リスク及び顧客に及ぼす潜在的な影響に基づく優先順位付けを用いて，組織は，ソフトウェア開発能力の自己評価の文書化した情報を保持しなければならない．

組織は，ソフトウェア開発を内部監査プログラム（9.2.2.1 参照）の範囲に含めなければならない．

gramme (see Section 9.2.2.1).

## 8.3.3 Design and development inputs

See ISO 9001:2015 requirements.

### 8.3.3.1 Product design input

The organization shall identify, document, and review product design input requirements as a result of contract review. Product design input requirements include but are not limited to the following:

a) product specifications including but not limited to special characteristics (see Section 8.3.3.3);

b) boundary and interface requirements;

c) identification, traceability, and packaging;

d) consideration of design alternatives;

e) assessment of risks with the input requirements and the organization's ability to mitigate/manage the risks, including from the feasibility analysis;

f) targets for conformity to product requirements including preservation, reliability, durability, serviceability, health, safety, envi-

> **8.3.3 設計・開発へのインプット**
> ISO 9001:2015 要求事項参照.

### 8.3.3.1 製品設計へのインプット

組織は，契約内容の確認の結果として，製品設計へのインプット要求事項を特定し，文書化し，レビューしなければならない．製品設計へのインプット要求事項には，次の事項を含める．しかし，それに限定されない．

a) 特殊特性（8.3.3.3 参照）を含む，しかし，それに限定されない製品仕様書

b) 境界及びインタフェース要求事項

c) 識別，トレーサビリティ及び包装

d) 設計の代替案の検討

e) インプット要求事項に伴うリスク及びリスクを緩和する／管理する組織の能力の，フィージビリティ分析の結果を含む評価

f) 保存，信頼性，耐久性，サービス性，健康，安全，環境，開発タイミング及びコストを含む，製品要求事項への適合に対する目標

ronmental, development timing, and cost;

g) applicable statutory and regulatory requirements of the customer-identified country of destination, if provided;

h) embedded software requirements.

The organization shall have a process to deploy information gained from previous design projects, competitive product analysis (benchmarking), supplier feedback, internal input, field data, and other relevant sources for current and future projects of a similar nature.

NOTE One approach for considering design alternatives is the use of trade-off curves.

**8.3.3.2 Manufacturing process design input**

The organization shall identify, document, and review manufacturing process design input requirements including but not limited to the following:

a) product design output data including special characteristics;

b) targets for productivity, process capability,

g)　顧客から提供された場合，顧客指定の仕向国の該当する法令・規制要求事項

h)　組込みソフトウェア要求事項

組織は，現在及び未来の類似するプロジェクトのために，過去の設計プロジェクト，競合製品分析(ベンチマーキング)，供給者からのフィードバック，内部からのインプット，市場データ及び他の関連する情報源から得られた情報を展開するプロセスをもたなければならない．

注記　設計の代替案を検討するアプローチの一つは，トレードオフ曲線の活用である．

### 8.3.3.2　製造工程設計へのインプット

組織は，製造工程設計へのインプット要求事項を特定し，文書化し，レビューしなければならない．これには次の事項を含めなければならない．しかし，それに限定されない．

a)　特殊特性を含む，製品設計からのアウトプットデータ

b)　生産性，工程能力，タイミング及びコストに対

timing, and cost;

c) manufacturing technology alternatives;
d) customer requirements, if any;
e) experience from previous developments;
f) new materials;
g) product handling and ergonomic requirements; and
h) design for manufacturing and design for assembly.

The manufacturing process design shall include the use of error-proofing methods to a degree appropriate to the magnitude of the problem(s) and commensurate with the risks encountered.

### 8.3.3.3 Special characteristics

The organization shall use a multidisciplinary approach to establish, document, and implement its process(es) to identify special characteristics, including those determined by the customer and the risk analysis performed by the organization, and shall include the following:

a) documentation of all special characteristics in the drawings (as required), risk analysis

する目標

c) 製造技術の代替案

d) もしあれば，顧客要求事項

e) 過去の開発からの経験

f) 新材料

g) 製品の取扱い及び人間工学的要求事項

h) 製造設計，組立設計

製造工程設計には，問題の大きさに対して適切な程度で，遭遇するリスクに見合う程度のポカヨケ手法の採用を含めなければならない．

**8.3.3.3 特殊特性**

組織は，顧客によって決定された，また組織によって実施されたリスク分析による特殊特性を特定するプロセスを確立し，文書化し，実施するために部門横断的アプローチを用いなければならない．それには次の事項を含めなければならない．

a) 図面（必要に応じて），リスク分析（FMEAのような），コントロールプラン及び標準作業／

(such as FMEA), control plans, and standard work/operator instructions; special characteristics are identified with specific markings and are cascaded through each of these documents;

b) development of control and monitoring strategies for special characteristics of products and production processes;

c) customer-specified approvals, when required;

d) compliance with customer-specified definitions and symbols or the organization's equivalent symbols or notations, as defined in a symbol conversion table. The symbol conversion table shall be submitted to the customer, if required.

### 8.3.4 Design and development controls

See ISO 9001:2015 requirements.

### 8.3.4.1 Monitoring

Measurements at specified stages during the design and development of products and processes shall be defined, analysed, and reported with summary results as an input to management re-

作業者指示書における全ての特殊特性の文書化．特殊特性は，固有の記号で識別され，これらの各文書を通じて展開される．

b)　製品及び生産工程の特殊特性に対する管理及び監視戦略の開発

c)　要求がある場合，顧客規定の承認

d)　顧客規定の定義及び記号，又は記号変換表に定められた，組織の同等の記号若しくは表記法への適合．記号変換表は，要求されれば顧客に提出しなければならない．

### 8.3.4　設計・開発の管理

ISO 9001:2015 要求事項参照．

#### 8.3.4.1　監視

製品及び工程の設計・開発中の規定された段階での測定項目を，定め，分析し，そして，その要約した結果をマネジメントレビューへのインプットとして報告しなければならない（9.3.2.1 参照）．

view (see Section 9.3.2.1).

When required by the customer, measurements of the product and process development activity shall be reported to the customer at stages specified, or agreed to, by the customer.

NOTE When appropriate, these measurements may include quality risks, costs, lead times, critical paths, and other measurements.

**8.3.4.2 Design and development validation**

Design and development validation shall be performed in accordance with customer requirements, including any applicable industry and governmental agency-issued regulatory standards. The timing of design and development validation shall be planned in alignment with customer-specified timing, as applicable.

Where contractually agreed with the customer, this shall include evaluation of the interaction of the organization's product, including embedded software, within the system of the final custom-

顧客に要求される場合，製品及び工程の開発活動の測定項目は，規定された段階で顧客に報告する，又は顧客に合意されなければならない．

注記　必要に応じて，測定項目には，品質リスク，コスト，リードタイム，クリティカルパス，などの測定項目を含めてもよい．

**8.3.4.2　設計・開発の妥当性確認**

設計・開発の妥当性確認は，該当する産業規格及び政府機関の発行する規制基準を含む，顧客要求事項に従って実行しなければならない．設計・開発の妥当性確認のタイミングは，該当する場合には，必ず，顧客規定のタイミングに合わせて計画しなければならない．

顧客との契約上の合意がある場合，設計・開発の妥当性確認には，顧客の最終製品のシステム内において，組込みソフトウェアを含めて，組織の製品の相互作用の評価を含めなければならない．

er's product.

#### 8.3.4.3 Prototype programme

When required by the customer, the organization shall have a prototype programme and control plan. The organization shall use, whenever possible, the same suppliers, tooling, and manufacturing processes as will be used in production.

All performance-testing activities shall be monitored for timely completion and conformity to requirements.

When services are outsourced, the organization shall include the type and extent of control in the scope of its quality management system to ensure that outsourced services conform to requirements (see ISO 9001, Section 8.4).

#### 8.3.4.4 Product approval process

The organization shall establish, implement, and maintain a product and manufacturing approval process conforming to requirements defined by the customer(s).

### 8.3.4.3　試作プログラム

顧客から要求される場合，組織は，試作プログラム及び試作コントロールプランをもたなければならない．組織は，可能な限り，量産で採用する同一の供給者，治工具及び製造工程を使用しなければならない．

タイムリーな完了及び要求事項への適合のために，全ての性能試験活動を監視しなければならない．

これらの業務をアウトソースする場合，組織は，アウトソースしたサービスが要求事項に適合することを確実にするために，管理の方式及び程度を品質マネジメントシステムの適用範囲に含めなければならない（ISO 9001 の 8.4 参照）．

### 8.3.4.4　製品承認プロセス

組織は，顧客に定められた要求事項に適合する，製品及び製造の承認プロセスを，確立し，実施し，維持しなければならない．

The organization shall approve externally provided products and services per ISO 9001, Section 8.4.3, prior to submission of their part approval to the customer.

The organization shall obtain documented product approval prior to shipment, if required by the customer. Records of such approval shall be retained.

NOTE Product approval should be subsequent to the verification of the manufacturing process.

### 8.3.5 Design and development outputs

See ISO 9001:2015 requirements.

### 8.3.5.1 Design and development outputs — supplemental

The product design output shall be expressed in terms that can be verified and validated against product design input requirements. The product design output shall include but is not limited to the following, as applicable:

組織は，自らの部品承認を顧客に提出するのに先立って，外部から提供される製品及びサービスをISO 9001の8.4.3によって承認しなければならない．

組織は，顧客に要求される場合，出荷に先立って，文書化した顧客の製品承認を取得しなければならない．そのような承認の記録は，保持しなければならない．

注記 製品承認は，製造工程が検証された後で実施することが望ましい．

### 8.3.5 設計・開発からのアウトプット

ISO 9001:2015要求事項参照．

#### 8.3.5.1 設計・開発からのアウトプット—補足

製品設計からのアウトプットは，製品設計へのインプット要求事項と対比した検証及び妥当性確認ができるように表現しなければならない．製品設計からのアウトプットには，該当する場合には，必ず，次の事項を含めなければならない．しかし，それに限定されない．

a) design risk analysis (FMEA);
b) reliability study results;
c) product special characteristics;
d) results of product design error-proofing, such as DFSS, DFMA, and FTA;
e) product definition including 3D models, technical data packages, product manufacturing information, and geometric dimensioning & tolerancing (GD&T);
f) 2D drawings, product manufacturing information, and geometric dimensioning & tolerancing (GD&T);
g) product design review results;
h) service diagnostic guidelines and repair and serviceability instructions;
i) service part requirements;
j) packaging and labeling requirements for shipping.

NOTE Interim design outputs should include any engineering problems being resolved through a trade-off process.

a)　設計リスク分析（FMEA）

b)　信頼性調査の結果

c)　製品の特殊特性

d)　DFSS，DFMA 及び FTA のような，製品設計のポカヨケの結果

e)　3D モデル，技術データパッケージ，製品製造の情報及び幾何寸法と公差（GD&T）を含む，製品の定義

f)　2D 図，製品製造の情報及び幾何寸法と公差（GD&T）

g)　製品デザインレビューの結果

h)　サービス故障診断の指針並びに修理及びサービス可能性の指示書

i)　サービス部品要求事項

j)　出荷のための包装及びラベリング要求事項

注記　暫定設計のアウトプットには，トレードオフプロセスを通じて解決された技術問題を含めることが望ましい．

### 8.3.5.2 Manufacturing process design output

The organization shall document the manufacturing process design output in a manner that enables verification against the manufacturing process design inputs. The organization shall verify the outputs against manufacturing process design input requirements. The manufacturing process design output shall include but is not limited to the following:

a) specifications and drawings;
b) special characteristics for product and manufacturing process;
c) identification of process input variables that impact characteristics;
d) tooling and equipment for production and control, including capability studies of equipment and process(es);
e) manufacturing process flow charts/layout, including linkage of product, process, and tooling;
f) capacity analysis;
g) manufacturing process FMEA;
h) maintenance plans and instructions;

### 8.3.5.2　製造工程設計からのアウトプット

組織は，製造工程設計からのアウトプットを，製造工程設計へのインプットと対比した検証ができるように文書化しなければならない．組織は，そのアウトプットを，製造工程設計へのインプット要求事項と対比して検証しなければならない．製造工程設計からのアウトプットには，次の事項を含めなければならない．しかし，それに限定されない．

a）仕様書及び図面
b）製品及び製造工程の特殊特性
c）特性に影響を与える，工程インプット変数の特定
d）設備及び工程の能力調査を含む，生産及び管理のための治工具及び設備
e）製品，工程及び治工具のつながりを含む，製造工程フローチャート／レイアウト
f）生産能力の分析
g）製造工程 FMEA
h）保全計画及び指示書

i) control plan (see Annex A);

j) standard work and work instructions;

k) process approval acceptance criteria;

l) data for quality, reliability, maintainability, and measurability;

m) results of error-proofing identification and verification, as appropriate;

n) methods of rapid detection, feedback, and correction of product/manufacturing process nonconformities.

### 8.3.6 Design and development changes

See ISO 9001:2015 requirements.

### 8.3.6.1 Design and development changes — supplemental

The organization shall evaluate all design changes after initial product approval, including those proposed by the organization or its suppliers, for potential impact on fit, form, function, performance, and/or durability. These changes shall be validated against customer requirements and approved internally, prior to production implementation.

i)　コントロールプラン（附属書 A 参照）

j)　標準作業及び作業指示書

k)　工程承認の合否判定基準

l)　品質，信頼性，保全性及び測定性に対するデータ

m)　必要に応じて，ポカヨケの特定及び検証の結果

n)　製品／製造工程の不適合の迅速な検出，フィードバック及び修正の方法

**8.3.6　設計・開発の変更**

ISO 9001:2015 要求事項参照．

**8.3.6.1　設計・開発の変更—補足**

組織は，組織又はその供給者から提案されたものを含めて，初回の製品承認の後の全ての設計変更を，取付時の合い，形状，機能，性能及び／又は耐久性に関する潜在的な影響に対して評価しなければならない．これらの変更は，生産で実施する前に，顧客要求事項に対する妥当性確認を実施して，内部で承認しなければならない．

If required by the customer, the organization shall obtain documented approval, or a documented waiver, from the customer prior to production implementation.

For products with embedded software, the organization shall document the revision level of software and hardware as part of the change record.

## 8.4 Control of externally provided processes, products and services

### 8.4.1 General

See ISO 9001:2015 requirements.

#### 8.4.1.1 General — supplemental

The organization shall include all products and services that affect customer requirements such as sub-assembly, sequencing, sorting, rework, and calibration services in the scope of their definition of externally provided products, processes, and services.

#### 8.4.1.2 Supplier selection process

The organization shall have a documented suppli-

顧客から要求される場合，組織は，文書化した承認，又は文書化した免除を，生産で実施する前に顧客から得なければならない．

組込みソフトウェアをもつ製品に対して，組織は，ソフトウェア及びハードウェアの改訂レベルを変更記録の一部として文書化しなければならない．

**8.4　外部から提供されるプロセス，製品及びサービスの管理**

**8.4.1　一般**

ISO 9001:2015 要求事項参照．

**8.4.1.1　一般—補足**

組織は，サブアセンブリ，整列，選別，手直し及び校正サービスのような，顧客要求事項に影響する全ての製品及びサービスを，外部から提供される製品，プロセス及びサービスの定義の範囲に含めなければならない．

**8.4.1.2　供給者選定プロセス**

組織は，文書化した供給者選定プロセスをもたな

er selection process. The selection process shall include:

a) an assessment of the selected supplier's risk to product conformity and uninterrupted supply of the organization's product to their customers;
b) relevant quality and delivery performance;
c) an evaluation of the supplier's quality management system;
d) multidisciplinary decision making; and
e) an assessment of software development capabilities, if applicable.

Other supplier selection criteria that should be considered include the following:

— volume of automotive business (absolute and as a percentage of total business);
— financial stability;
— purchased product, material, or service complexity;
— required technology (product or process);
— adequacy of available resources (e.g., people, infrastructure);
— design and development capabilities (includ-

ければならない．選定プロセスには，次の事項を含めなければならない．

a)　選定される供給者の製品適合性及び顧客に対する組織の製品の途切れない供給に対するリスクの評価

b)　関連する品質及び納入パフォーマンス

c)　供給者の品質マネジメントシステムの評価

d)　部門横断的意思決定

e)　該当する場合には，必ず，ソフトウェア開発能力の評価

ほかにも供給者の選定基準には，次の事項を考慮することが望ましい．

—　自動車事業の規模（絶対値及び事業全体における割合）

—　財務的安定性

—　購入される製品，材料，又はサービスの複雑さ

—　必要な技術（製品又はプロセス）

—　利用可能な資源（例　人材，インフラストラクチャ）の適切性

—　設計・開発の能力（プロジェクトマネジメント

ing project management);

— manufacturing capability;
— change management process;
— business continuity planning (e.g., disaster preparedness, contingency planning);
— logistics process;
— customer service.

#### 8.4.1.3 Customer-directed sources (also known as "Directed-Buy")

When specified by the customer, the organization shall purchase products, materials, or services from customer-directed sources.

All requirements of Section 8.4 (except the requirements in IATF 16949, Section 8.4.1.2) are applicable to the organization's control of customer-directed sources unless specific agreements are otherwise defined by the contract between the organization and the customer.

### 8.4.2 Type and extent of control

See ISO 9001:2015 requirements.

を含む．）

— 製造の能力
— 変更管理プロセス
— 事業継続計画（例　災害への準備，緊急事態対応計画）
— 物流プロセス
— 顧客サービス

**8.4.1.3　顧客指定の供給者（“指定購買”としても知られる）**

顧客に規定された場合，組織は，製品，材料，又はサービスを顧客指定の供給者から購買しなければならない．

8.4 の全ての要求事項（IATF 16949 の 8.4.1.2 の要求事項を除く）は，組織と顧客との間で契約によって定められた特定の合意がない限り，組織の顧客指定の供給者の管理に対して，適用される．

**8.4.2　管理の方式及び程度**

ISO 9001:2015 要求事項参照．

### 8.4.2.1 Type and extent of control — supplemental

The organization shall have a documented process to identify outsourced processes and to select the types and extent of controls used to verify conformity of externally provided products, processes, and services to internal (organizational) and external customer requirements.

The process shall include the criteria and actions to escalate or reduce the types and extent of controls and development activities based on supplier performance and assessment of product, material, or service risks.

### 8.4.2.2 Statutory and regulatory requirements

The organization shall document their process to ensure that purchased products, processes, and services conform to the current applicable statutory and regulatory requirements in the country of receipt, the country of shipment, and the customer-identified country of destination, if provided.

#### 8.4.2.1　管理の方式及び程度—補足

組織は，アウトソースしたプロセスを特定する，並びに，外部から提供される製品，プロセス及びサービスに対し，内部（組織）及び外部顧客の要求事項への適合を検証するために用いる管理の方式及び程度を選定する，文書化したプロセスをもたなければならない．

そのプロセスには，管理の方式及び程度を拡大する又は縮小する判断基準及び処置，並びに供給者のパフォーマンス，及び製品，材料又はサービスのリスクの評価に基づく開発活動を含めなければならない．

#### 8.4.2.2　法令・規制要求事項

組織は，購入した製品，プロセス及びサービスが，受入国，出荷国及び顧客に特定された仕向国の現在該当する法令・規制要求事項が提供されれば，その要求事項に適合することを確実にするプロセスを文書化しなければならない．

If the customer defines special controls for certain products with statutory and regulatory requirements, the organization shall ensure they are implemented and maintained as defined, including at suppliers.

### 8.4.2.3 Supplier quality management system development

The organization shall require their suppliers of automotive products and services to develop, implement, and improve a quality management system certified to ISO 9001, unless otherwise authorized by the customer [e.g., item a) below], with the ultimate objective of becoming certified to this Automotive QMS Standard. Unless otherwise specified by the customer, the following sequence should be applied to achieve this requirement:

a) compliance to ISO 9001 through second-party audits;

b) certification to ISO 9001 through third-party audits; unless otherwise specified by the customer, suppliers to the organization shall demonstrate conformity to ISO 9001 by maintaining a third-party certification issued by

顧客が，法令・規制要求事項をもつ製品に対して特別管理を定めているならば，組織は，供給者で管理する場合を含めて，定められたとおりに実施し，維持することを確実にしなければならない．

#### 8.4.2.3 供給者の品質マネジメントシステム開発

組織は，自動車の製品及びサービスの供給者に，顧客による他の許可［例　下記のa)］がない限り，この自動車産業QMS規格に認証されることを最終的な目標として，ISO 9001に認証された品質マネジメントシステムの開発，実施及び改善を要求しなければならない．この要求事項を達成するために，次の順序を適用することが望ましい．ただし，顧客によって他に規定されたときは，この限りではない．

a) 第二者監査を通じたISO 9001に対する適合

b) 第三者審査を通じたISO 9001に対する認証．顧客による他の規定がない限り，組織への供給者はISO 9001に対する認証を実証しなければならない．実証するには認定機関の主要適用範囲がISO/IEC 17021へのマネジメント

a certification body bearing the accreditation mark of a recognized IAF MLA (International Accreditation Forum Multilateral Recognition Arrangement) member and where the accreditation body's main scope includes management system certification to ISO/IEC 17021;

c) certification to ISO 9001 with compliance to other customer-defined QMS requirements (such as Minimum Automotive Quality Management System Requirements for Sub-Tier Suppliers [MAQMSR] or equivalent) through second-party audits;

d) certification to ISO 9001 with compliance to IATF 16949 through second-party audits;

e) certification to 16949 through third-party audits (valid third-party certification of the supplier to IATF 16949 by an IATF-recognized certification body).

#### 8.4.2.3.1 Automotive product-related software or automotive products with embedded software

The organization shall require their suppliers of

システム認証を含む場合の，承認されたIAF MLA（International Accreditation Forum Multilateral Recognition Arrangement）メンバーの認定マークをもつ認証機関が発行する第三者認証を維持する．

c）第二者監査を通じた，顧客が定めた他のQMS要求事項（例えばMinimum Automotive Quality Management System Requirements for Sub-Tier Suppliers［MAQMSR］又はそれに相当するもの）への適合を伴うISO 9001に対する認証
d）第二者監査を通じたIATF 16949に対する適合を伴うISO 9001への認証
e）第三者審査を通じた16949に対する認証（IATFが認めた認証機関による，IATF 16949への供給者の有効な第三者認証）

**8.4.2.3.1　自動車製品に関係するソフトウェア又は組込みソフトウェアをもつ製品**

組織は，自動車製品に関係するソフトウェアの供

automotive product-related software, or automotive products with embedded software, to implement and maintain a process for software quality assurance for their products.

A software development assessment methodology shall be utilized to assess the supplier's software development process. Using prioritization based on risk and potential impact to the customer, the organization shall require the supplier to retain documented information of a software development capability self-assessment.

### 8.4.2.4 Supplier monitoring

The organization shall have a documented process and criteria to evaluate supplier performance in order to ensure conformity of externally provided products, processes, and services to internal and external customer requirements.

At a minimum, the following supplier performance indicators shall be monitored:

a) delivered product conformity to requirements;
b) customer disruptions at the receiving plant,

給者，又は組込みソフトウェアをもつ自動車製品の供給者に，その製品に対するソフトウェア品質保証のためのプロセスを実施し維持することを要求しなければならない．

ソフトウェア開発評価の方法論は，供給者のソフトウェア開発を評価するために活用しなければならない．リスク及び顧客へ及ぼす潜在的影響に基づく優先順位付けを用いて，組織は，供給者にソフトウェア開発能力の自己評価の文書化した情報を保持するよう要求しなければならない．

#### 8.4.2.4　供給者の監視

組織は，外部から提供される製品，プロセス及びサービスの内部及び外部顧客の要求事項への適合を確実にするために，供給者のパフォーマンスを評価する，文書化したプロセス及び判断基準をもたなければならない．

最低限，次の供給者のパフォーマンス指標を監視しなければならない．

a)　納入された製品の要求事項への適合

b)　構内保留及び出荷停止を含む，受入工場におい

including yard holds and stop ships;

c) delivery schedule performance;

d) number of occurrences of premium freight.

If provided by the customer, the organization shall also include the following, as appropriate, in their supplier performance monitoring:

e) special status customer notifications related to quality or delivery issues;

f) dealer returns, warranty, field actions, and recalls.

### 8.4.2.4.1 Second-party audits

The organization shall include a second-party audit process in their supplier management approach. Second-party audits may be used for the following:

a) supplier risk assessment;

b) supplier monitoring;

c) supplier QMS development;

d) product audits;

e) process audits.

Based on a risk analysis, including product safety/

て顧客が被った迷惑

c)　納期パフォーマンス

d)　特別輸送費の発生件数

もし顧客から提供されれば，組織は，必要に応じて，次の事項も供給者パフォーマンスの監視に含めなければならない．

e)　品質問題又は納期問題に関係する，特別状態の顧客からの通知

f)　ディーラーからの返却，補償，市場処置及びリコール

### 8.4.2.4.1　第二者監査

組織は，供給者の管理方法に第二者監査プロセスを含めなければならない．第二者監査は，次の事項に対して使用してもよい．

a)　供給者のリスク評価

b)　供給者の監視

c)　供給者の QMS 開発

d)　製品監査

e)　工程監査

最低限，製品安全／規制要求事項，供給者のパフ

regulatory requirements, performance of the supplier, and QMS certification level, at a minimum, the organization shall document the criteria for determining the need, type, frequency, and scope of second-party audits.

The organization shall retain records of the second-party audit reports.

If the scope of the second-party audit is to assess the supplier's quality management system, then the approach shall be consistent with the automotive process approach.

NOTE Guidance may be found in the IATF Auditor Guide and ISO 19011.

#### 8.4.2.5 Supplier development

The organization shall determine the priority, type, extent, and timing of required supplier development actions for its active suppliers. Determination inputs shall include but are not limited to the following:

a) performance issues identified through suppli-

ォーマンス及びQMS認証レベルを含む，リスク分析に基づいて，組織は，第二者監査の必要性，方式，頻度及び範囲を決定するための基準を文書化しなければならない．

組織は，第二者監査報告書の記録を保持しなければならない．

第二者監査の範囲が供給者の品質マネジメントシステムを評価する場合，その方法は自動車産業プロセスアプローチと整合性がとれていなければならない．

注記　手引は，IATF審査員ガイド及びISO 19011に見いだすことができる．

### 8.4.2.5　供給者の開発

組織は，現行の供給者に対し，必要な供給者開発の優先順位，方式，程度及びタイミングを決定しなければならない．決定をするためのインプットには次の事項を含めなければならない．しかし，それに限定されない．

a)　供給者の監視（8.4.2.4参照）を通じて特定さ

er monitoring (see Section 8.4.2.4);

b) second-party audit findings (see Section 8.4.2.4.1);

c) third-party quality management system certification status;

d) risk analysis.

The organization shall implement actions necessary to resolve open (unsatisfactory) performance issues and pursue opportunities for continual improvement.

**8.4.3 Information for external providers**

See ISO 9001:2015 requirements.

**8.4.3.1 Information for external providers — supplemental**

The organization shall pass down all applicable statutory and regulatory requirements and special product and process characteristics to their suppliers and require the suppliers to cascade all applicable requirements down the supply chain to the point of manufacture.

れたパフォーマンス問題

b)　第二者監査の所見（8.4.2.4.1 参照）

c)　第三者品質マネジメントシステム認証の状態

d)　リスク分析

組織は，未解決（未達）のパフォーマンス問題を解決するため及び継続的改善に対する機会を追求するために必要な処置を実施しなければならない．

### 8.4.3　外部提供者に対する情報

ISO 9001:2015 要求事項参照．

### 8.4.3.1　外部提供者に対する情報―補足

組織は，全ての該当する法令・規制要求事項，並びに製品及び工程の特殊特性を供給者に引き渡し，サプライチェーンをたどって，製造現場にまで，全ての該当する要求事項を展開するよう，供給者に要求しなければならない．

## 8.5 Production and service provision

### 8.5.1 Control of production and service provision

See ISO 9001:2015 requirements.

NOTE Suitable infrastructure includes appropriate manufacturing equipment required to ensure product compliance. Monitoring and measuring resources include appropriate monitoring and measuring equipment required to ensure effective control of manufacturing processes.

#### 8.5.1.1 Control plan

The organization shall develop control plans (in accordance with Annex A) at the system, subsystem, component, and/or material level for the relevant manufacturing site and all product supplied, including those for processes producing bulk materials as well as parts. Family control plans are acceptable for bulk material and similar parts using a common manufacturing process.

The organization shall have a control plan for pre-

## 8.5 製造及びサービス提供

### 8.5.1 製造及びサービス提供の管理

ISO 9001:2015 要求事項参照.

注記 適切なインフラストラクチャは，製品の適合を確実にするために必要な，適切な製造設備を含む．監視及び測定のための資源は，製造工程の効果的な管理を確実にするために必要な，適切な監視及び測定設備を含む．

#### 8.5.1.1 コントロールプラン

組織は，該当する製造サイト及び全ての供給する製品に対して，コントロールプラン（附属書 A に従って）を，システム，サブシステム，構成部品及び／又は材料のレベルで，部品だけではなくバルク材料を含めて，策定しなければならない．ファミリーコントロールプランは，バルク材料及び共通の製造工程を使う類似の部品に対して容認される．

組織は，量産試作及び量産に対して，どのように

launch and production that shows linkage and incorporates information from the design risk analysis (if provided by the customer), process flow diagram, and manufacturing process risk analysis outputs (such as FMEA).

The organization shall, if required by the customer, provide measurement and conformity data collected during execution of either the pre-launch or production control plans. The organization shall include in the control plan:

a) controls used for the manufacturing process control, including verification of job set-ups;
b) first-off/last-off part validation, as applicable;

c) methods for monitoring of control exercised over special characteristics (see Annex A) defined by both the customer and the organization;
d) the customer-required information, if any;
e) specified reaction plan (see Annex A); when nonconforming product is detected, the process becomes statistically unstable or not statistically capable.

つながっているかを示し（もし顧客から提供されれば）設計リスク分析からの情報や，工程フロー図及び製造工程のリスク分析のアウトプット（FMEAのような）からの情報を反映する，コントロールプランをもたなければならない．

組織は，顧客から要求される場合，量産試作又は量産コントロールプランを実行したときに集めた測定及び適合データを顧客に提供しなければならない．組織は，次の事項をコントロールプランに含めなければならない．

a)　作業の段取り替え検証を含む，製造工程の管理に使用される管理手段

b)　該当する場合には，必ず，初品／終品の妥当性確認

c)　顧客及び組織の双方で定められた，特殊特性に施される管理の監視方法（附属書A参照）

d)　もしあれば，顧客から要求される情報

e)　不適合製品が検出された場合，工程が統計的に不安定又は統計的に能力不足になった場合の，規定された対応計画（附属書A参照）．

The organization shall review control plans, and update as required, for any of the following:

f) the organization determines it has shipped nonconforming product to the customer;
g) when any change occurs affecting product, manufacturing process, measurement, logistics, supply sources, production volume changes, or risk analysis (FMEA) (see Annex A);
h) after a customer complaint and implementation of the associated corrective action, when applicable;
i) at a set frequency based on a risk analysis.

If required by the customer, the organization shall obtain customer approval after review or revision of the control plan.

### 8.5.1.2 Standardised work — operator instructions and visual standards

The organization shall ensure that standardised work documents are:

a) communicated to and understood by the employees who are responsible for performing

組織は，次の事項が発生した場合，コントロールプランをレビューし，必要に応じて更新しなければならない．

f)　不適合製品を顧客に出荷したと組織が判断した場合

g)　製品，製造工程，測定，物流，供給元，生産量変更，又はリスク分析（FMEA）に影響する，変更が発生した場合（附属書 A 参照）

h)　該当する場合には，必ず，顧客苦情及び関連する是正処置が実施された後

i)　リスク分析に基づく，設定された頻度で

顧客に要求されれば，組織は，コントロールプランのレビュー又は改訂の後で，顧客の承認を得なければならない．

### 8.5.1.2　標準作業—作業者指示書及び目視標準

組織は，標準作業文書が次のとおりであることを確実にしなければならない．

a)　作業を行う責任をもつ従業員に伝達され，理解される．

the work;

b) legible;

c) presented in the language(s) understood by the personnel responsible to follow them;

d) accessible for use at the designated work area(s).

The standardised work documents shall also include rules for operator safety.

### 8.5.1.3 Verification of job set-ups

The organization shall:

a) verify job set-ups when performed, such as an initial run of a job, material changeover, or job change that requires a new set-up;

b) maintain documented information for set-up personnel;

c) use statistical methods of verification, where applicable;

d) perform first-off/last-off part validation, as applicable; where appropriate, first-off parts should be retained for comparison with the last-off parts; where appropriate, last-off-

b)　読みやすい．
c)　それに従う責任のある要員に理解される言語で提供する．
d)　指定された作業現場で利用可能である．

標準作業文書には，作業者の安全に対する規則も，含めなければならない．

### 8.5.1.3　作業の段取り替え検証

組織は，次の事項を実施しなければならない．

a)　作業の立上げ，材料切替え，又は作業変更のような新しい段取り替えが実行される場合は，作業の段取り替えを検証する．

b)　段取り替え要員のために文書化した情報を維持する．
c)　該当する場合には，必ず，検証に統計的方法を使用する．
d)　該当する場合には，必ず，初品／終品の妥当性確認を実施する．必要に応じて，初品は終品との比較のために保持し，終品は次の工程稼働まで保持することが望ましい．

parts should be retained for comparison with first-off parts in subsequent runs;

e) retain records of process and product approval following set-up and first-off/last-off part validations.

### 8.5.1.4 Verification after shutdown

The organization shall define and implement the necessary actions to ensure product compliance with requirements after a planned or unplanned production shutdown period.

### 8.5.1.5 Total productive maintenance

The organization shall develop, implement, and maintain a documented total productive maintenance system.

At a minimum, the system shall include the following:

a) identification of process equipment necessary to produce conforming product at the required volume;

b) availability of replacement parts for the

e)　段取り替え及び初品／終品の妥当性確認後の工程及び製品承認の記録を保持する.

### 8.5.1.4　シャットダウン後の検証

組織は，計画的又は非計画的シャットダウン後に，製品が要求事項に適合することを確実にするのに必要な処置を定め，実施しなければならない.

### 8.5.1.5　TPM（Total productive maintenance）

組織は，文書化したTPMシステムを構築し，実施し，維持しなければならない.

そのシステムには，最低限，次の事項を含めなければならない.

a)　要求された量の適合製品を生産するために必要な工程設備の特定

b)　a)で特定された設備に対する交換部品の入手

equipment identified in item a);

c) provision of resource for machine, equipment, and facility maintenance;

d) packaging and preservation of equipment, tooling, and gauging;

e) applicable customer-specific requirements;

f) documented maintenance objectives, for example: OEE (Overall Equipment Effectiveness), MTBF (Mean Time Between Failure), and MTTR (Mean Time To Repair), and Preventive Maintenance compliance metrics. Performance to the maintenance objectives shall form an input into management review (see ISO 9001, Section 9.3);

g) regular review of maintenance plan and objectives and a documented action plan to address corrective actions where objectives are not achieved;

h) use of preventive maintenance methods;

i) use of predictive maintenance methods, as applicable;

j) periodic overhaul.

性

c)　機械，設備及び施設の保全のための資源の提供

d)　設備，治工具及びゲージの包装及び保存

e)　該当する顧客固有要求事項

f)　文書化した保全目標，例えば，OEE（総合設備効率），MTBF（平均故障間隔）及びMTTR（平均修理時間），並びに予防保全の順守指標．保全目標に対するパフォーマンスは，マネジメントレビュー（ISO 9001 の 9.3 参照）へのインプットとしなければならない．

g)　目標が未達であった場合の，保全計画及び目標，並びに是正処置に取り組む文書化した処置計画に関する定期的レビュー

h)　予防保全の方法の使用

i)　該当する場合には，必ず，予知保全の方法の使用

j)　定期的オーバーホール

## 8.5.1.6 Management of production tooling and manufacturing, test, inspection tooling and equipment

The organization shall provide resources for tool and gauge design, fabrication, and verification activities for production and service materials and for bulk materials, as applicable.

The organization shall establish and implement a system for production tooling management, whether owned by the organization or the customer, including:

a) maintenance and repair facilities and personnel;
b) storage and recovery;
c) set-up;
d) tool-change programmes for perishable tools;
e) tool design modification documentation, including engineering change level of the product;
f) tool modification and revision to documentation;
g) tool identification, such as serial or asset number; the status, such as production, re-

### 8.5.1.6　生産治工具並びに製造，試験，検査の治工具及び設備の運用管理

組織は，該当する場合には，必ず，生産及びサービス用材料並びにバルク材のための治工具及びゲージの設計，製作，及び検証活動に対して資源を提供しなければならない．

組織は，次の事項を含む，組織所有又は顧客所有の生産治工具であっても，これらの運用管理システムを確立し，実施しなければならない．

a)　保全及び修理用施設並びに要員

b)　保管及び補充

c)　段取り替え

d)　消耗する治工具の交換プログラム

e)　製品の技術変更レベルを含む，治工具設計変更の文書化

f)　治工具の改修及び文書の改訂

g)　シリアル番号又は資産番号のような，生産中，修理中又は廃却のような状況，所有者及び場所

pair or disposal; ownership; and location.

The organization shall verify that customer-owned tools, manufacturing equipment, and test/inspection equipment are permanently marked in a visible location so that the ownership and application of each item can be determined.

The organization shall implement a system to monitor these activities if any work is outsourced.

### 8.5.1.7 Production scheduling

The organization shall ensure that production is scheduled in order to meet customer orders/demands such as Just-In-Time (JIT) and is supported by an information system that permits access to production information at key stages of the process and is order driven.

The organization shall include relevant planning information during production scheduling, e.g., customer orders, supplier on-time delivery performance, capacity, shared loading (multi-part sta-

に関する治工具の識別

組織は，顧客所有の治工具，製造設備及び試験／検査設備に，所有権及び各品目の適用が明確になるように，見やすい位置に恒久的マークが付いていることを検証しなければならない．

組織は，作業がアウトソースされる場合，これらの活動を監視するシステムを実施しなければならない．

#### 8.5.1.7　生産計画

組織は，ジャストインタイム（JIT）のような顧客の注文／需要を満たすように生産を計画すること，及び，キーとなる工程の生産情報にアクセスでき受注生産方式に対応した情報システムによって生産を支援することを確実にしなければならない．

組織は，生産計画中に，関連する計画情報，例えば，顧客注文，供給者オンタイム納入パフォーマンス，生産能力，共通の負荷（複数部品加工場），リードタイム，在庫レベル，予防保全及び校正を含め

tion), lead time, inventory level, preventive maintenance, and calibration.

## 8.5.2 Identification and traceability

See ISO 9001:2015 requirements.

NOTE Inspection and test status is not indicated by the location of product in the production flow unless inherently obvious, such as material in an automated production transfer process. Alternatives are permitted if the status is clearly identified, documented, and achieves the designated purpose.

### 8.5.2.1 Identification and traceability — supplemental

The purpose of traceability is to support identification of clear start and stop points for product received by the customer or in the field that may contain quality and/or safety-related nonconformities. Therefore, the organization shall implement identification and traceability processes as described below.

なければならない.

#### 8.5.2　識別及びトレーサビリティ

ISO 9001:2015 要求事項参照.

注記　検査及び試験の状態は，自動化された製造搬送工程中の材料のように本質的に明確である場合を除き，生産フローにおける製品の位置によっては示されない．状態が明確に識別され，文書化され，規定された目的を達成するならば，代替手段が認められる.

#### 8.5.2.1　識別及びトレーサビリティ―補足

トレーサビリティの目的は，顧客が受け入れた製品，又は市場において品質及び／又は安全関係の不適合を含んでいる可能性がある製品に対して，開始，停止時点を明確に特定することを支援するためにある．したがって，組織は，識別及びトレーサビリティのプロセスを下記に記載されているとおりに実施しなければならない.

The organization shall conduct an analysis of internal, customer, and regulatory traceability requirements for all automotive products, including developing and documenting traceability plans, based on the levels of risk or failure severity for employees, customers, and consumers. These plans shall define the appropriate traceability systems, processes, and methods by product, process, and manufacturing location that:

a) enable the organization to identify nonconforming and/or suspect product;
b) enable the organization to segregate nonconforming and/or suspect product;
c) ensure the ability to meet the customer and/or regulatory response time requirements;
d) ensure documented information is retained in the format (electronic, hardcopy, archive) that enables the organization to meet the response time requirements;
e) ensure serialized identification of individual products, if specified by the customer or regulatory standards;
f) ensure the identification and traceability requirements are extended to externally provid-

組織は，全ての自動車製品に対して，従業員，顧客及び消費者に対するリスクのレベル又は故障の重大性に基づいて，トレーサビリティ計画の策定及び文書化を含めて，内部，顧客及び規制のトレーサビリティ要求事項の分析を実施しなければならない．その計画は，製品，プロセス及び製造場所ごとに，適切なトレーサビリティシステム，プロセス及び方法を，次のようになるように，定めなければならない．

a)　組織が，不適合製品及び／又は疑わしい製品を識別できるようにする．
b)　組織が，不適合製品及び／又は疑わしい製品を分別できるようにする．
c)　顧客及び／又は規制の対応時間の要求事項を満たす能力を確実にする．
d)　組織が対応時間の要求事項を満たせるようにできる様式（電子版，印刷版，保管用）で文書化した情報を保持することを確実にする．
e)　顧客又は規制基準によって規定されている場合，個別製品のシリアル化された識別を確実にする．
f)　識別及びトレーサビリティ要求事項が，安全／規制特性をもつ，外部から提供される製品に拡

ed products with safety/regulatory characteristics.

### 8.5.3 Property belonging to customers or external providers

See ISO 9001:2015 requirements.

### 8.5.4 Preservation

See ISO 9001:2015 requirements.

### 8.5.4.1 Preservation — supplemental

Preservation shall include identification, handling, contamination control, packaging, storage, transmission or transportation, and protection.

Preservation shall apply to materials and components from external and/or internal providers from receipt through processing, including shipment and until delivery to/acceptance by the customer.

In order to detect deterioration, the organization shall assess at appropriate planned intervals the condition of product in stock, the place/type of

張適用することを確実にする.

#### 8.5.3　顧客又は外部提供者の所有物

ISO 9001:2015 要求事項参照.

#### 8.5.4　保存

ISO 9001:2015 要求事項参照.

#### 8.5.4.1　保存―補足

保存に関わる考慮事項には，識別，取扱い，汚染防止，包装，保管，伝送又は輸送及び保護を含めなければならない.

保存は，外部及び／又は内部の提供者からの材料及び構成部品に，受領から加工を通じて，顧客への納入／顧客による受入れまでを含めて，適用しなければならない.

劣化を検出するために，組織は，保管中の製品の状態，保管容器の場所／方式及び保管環境を，適切に予定された間隔で評価しなければならない.

storage container, and the storage environment.

The organization shall use an inventory management system to optimize inventory turns over time and ensure stock rotation, such as "first-in-first-out" (FIFO).

The organization shall ensure that obsolete product is controlled in a manner similar to that of nonconforming product.

Organizations shall comply with preservation, packaging, shipping, and labeling requirements as provided by their customers.

### 8.5.5 Post-delivery activities

See ISO 9001:2015 requirements.

### 8.5.5.1 Feedback of information from service

The organization shall ensure that a process for communication of information on service concerns to manufacturing, material handling, logistics, engineering, and design activities is established,

組織は，在庫回転時間を最適化するため及び“先入れ先出し”（FIFO）のような，在庫の回転を確実にするために，在庫管理システムを使用しなければならない．

組織は，旧式となった製品は，不適合製品と同様な方法で管理することを，確実にしなければならない．

組織は，顧客から提供された，保存，包装，出荷及びラベリング要求事項に適合しなければならない．

### 8.5.5　引渡し後の活動

ISO 9001:2015 要求事項参照．

#### 8.5.5.1　サービスからの情報のフィードバック

組織は，製造，材料の取扱い，物流，技術及び設計活動へのサービスの懸念事項に関する情報を伝達するプロセスを確立し，実施し，維持することを確実にしなければならない．

implemented, and maintained.

NOTE 1 The intent of the addition of "service concerns" to this sub-clause is to ensure that the organization is aware of nonconforming product(s) and material(s) that may be identified at the customer location or in the field.

NOTE 2 "Service concerns" should include the results of field failure test analysis (see Section 10.2.6) where applicable.

#### 8.5.5.2 Service agreement with customer

When there is a service agreement with the customer, the organization shall:

a) verify that the relevant service centres comply with applicable requirements;
b) verify the effectiveness of any special purpose tools or measurement equipment;
c) ensure that all service personnel are trained in applicable requirements.

| **8.5.6 Control of changes** |
| --- |
| See ISO 9001:2015 requirements. |

注記 1　この箇条に“サービスの懸念事項”を追加する意図は，顧客のサイト又は市場で特定される可能性がある，不適合製品及び不適合材料を組織が認識することを確実にするためである．

注記 2　“サービスの懸念事項”に，該当する場合には，必ず，市場不具合の試験解析（10.2.6 参照）の結果を含めることが望ましい．

#### 8.5.5.2　顧客とのサービス契約

顧客とのサービス契約がある場合，組織は次の事項を実施しなければならない．

a)　関連するサービスセンターが，該当する要求事項に適合することを検証する．

b)　特殊目的治工具又は測定設備の有効性を検証する．

c)　全てのサービス要員が該当する要求事項について教育訓練されていることを確実にする．

### 8.5.6　変更の管理

ISO 9001:2015 要求事項参照．

### 8.5.6.1 Control of changes — supplemental

The organization shall have a documented process to control and react to changes that impact product realization. The effects of any change, including those changes caused by the organization, the customer, or any supplier, shall be assessed.

The organization shall:

a) define verification and validation activities to ensure compliance with customer requirements;
b) validate changes before implementation;
c) document the evidence of related risk analysis;
d) retain records of verification and validation.

Changes, including those made at suppliers, should require a production trial run for verification of changes (such as changes to part design, manufacturing location, or manufacturing process) to validate the impact of any changes on the manufacturing process.

When required by the customer, the organization

### 8.5.6.1 変更の管理―補足

組織は，製品実現に影響する変更を管理し対応する，文書化したプロセスをもたなければならない．組織，顧客，又は供給者に起因する変更を含む，変更の影響を評価しなければならない．

組織は，次の事項を実施しなければならない．

a) 顧客要求事項への適合を確実にするための検証及び妥当性確認の活動を定める．

b) 実施の前に変更の妥当性確認を行う．

c) 関係するリスク分析の証拠を文書化する．

d) 検証及び妥当性確認の記録を保持する．

供給者で行う変更を含めて，変更は，製造工程に与える変更の影響の妥当性確認を行うために，その変更点（部品設計，製造場所，又は製造工程の変更のような）の検証に対する生産トライアル稼働を要求することが望ましい．

顧客に要求される場合，組織は，次の事項を実施

shall:

e) notify the customer of any planned product realization changes after the most recent product approval;

f) obtain documented approval, prior to implementation of the change;

g) complete additional verification or identification requirements, such as production trial run and new product validation.

**8.5.6.1.1 Temporary change of process controls**

The organization shall identify, document, and maintain a list of the process controls, including inspection, measuring, test, and error-proofing devices, that includes the primary process control and the approved back-up or alternate methods.

The organization shall document the process that manages the use of alternate control methods. The organization shall include in this process, based on risk analysis (such as FMEA), severity, and the internal approvals to be obtained prior to production implementation of the alternate control

しなければならない．

e)　直近の製品承認の後の，計画した製品実現の変更を顧客に通知する．

f)　変更の実施の前に，文書化した承認を得る．

g)　生産トライアル稼働及び新製品の妥当性確認のような，追加の検証又は識別の要求事項を完了する．

#### 8.5.6.1.1　工程管理の一時的変更

組織は，検査，測定，試験及びポカヨケ装置を含む，工程管理のリストを特定し，文書化し，維持しなければならない．そのリストには，当初の工程管理及び承認されたバックアップ又は代替方法を含める．

組織は，代替管理方法の使用を運用管理するプロセスを文書化しなければならない．組織は，このプロセスに，リスク分析（FMEAのような）に基づいて，重大性及び代替管理方法の生産実施の前に取得する内部承認を含めなければならない．

method.

Before shipping product that was inspected or tested using the alternate method, if required, the organization shall obtain approval from the customer(s). The organization shall maintain and periodically review a list of approved alternate process control methods that are referenced in the control plan.

Standard work instructions shall be available for each alternate process control method. The organization shall review the operation of alternate process controls on a daily basis, at a minimum, to verify implementation of standard work with the goal to return to the standard process as defined by the control plan as soon as possible. Example methods include but are not limited to the following:

a) daily quality focused audits (e.g., layered process audits, as applicable);
b) daily leadership meetings.

Restart verification is documented for a defined

代替手法を使用して，検査され又は試験された製品の出荷の前に，要求される場合，組織は，顧客の承認を取得しなければならない．組織は，コントロールプランに引用され，承認された代替工程管理方法のリストを，維持し，定期的にレビューしなければならない．

標準作業指示書は，各代替工程管理方法に対して利用可能でなければならない．組織は，コントロールプランに定められた標準工程に可及的速やかに復帰することを目標とする，標準作業の実施を検証するために，代替工程管理の運用を，最低限，日常的にレビューしなければならない．方法例には次の事項を含める．しかし，それに限定されない．

a) 日常的品質重視監査（例　該当する場合には，必ず，階層別工程監査）
b) 日常的リーダーシップ会議

重大性によって定められている期間及びポカヨケ

period based on severity and confirmation that all features of the error-proofing device or process are effectively reinstated.

The organization shall implement traceability of all product produced while any alternate process control devices or processes are being used (e.g., verification and retention of first piece and last piece from every shift).

## 8.6 Release of products and services

See ISO 9001:2015 requirements.

### 8.6.1 Release of products and services — supplemental

The organization shall ensure that the planned arrangements to verify that the product and service requirements have been met encompass the control plan and are documented as specified in the control plan (see Annex A).

The organization shall ensure that the planned arrangements for initial release of products and services encompass product or service approval.

装置又は工程の全ての機能が有効に復帰しているとの確認に基づいて，再稼働検証を文書化する.

組織は，代替工程管理装置又は工程が使用されていた間，生産された全ての製品に対しトレーサビリティを実施しなければならない（例　シフトごとに得られた初品及び終品の検証及び保管）.

## 8.6 製品及びサービスのリリース

ISO 9001:2015 要求事項参照.

### 8.6.1 製品及びサービスのリリース—補足

組織は，製品及びサービスの要求事項が満たされていることを検証するための計画した取決めが，コントロールプランを網羅し，かつ，コントロールプラン（附属書 A 参照）に規定されたように文書化されていることを確実にしなければならない.

組織は，製品及びサービスの初回リリースに対する計画した取決めが，製品又はサービスの承認を網羅することを確実にしなければならない.

The organization shall ensure that product or service approval is accomplished after changes following initial release, according to ISO 9001, Section 8.5.6.

### 8.6.2 Layout inspection and functional testing

A layout inspection and a functional verification to applicable customer engineering material and performance standards shall be performed for each product as specified in the control plans. Results shall be available for customer review.

NOTE 1 Layout inspection is the complete measurement of all product dimensions shown on the design record(s).
NOTE 2 The frequency of layout inspection is determined by the customer.

### 8.6.3 Appearance items

For organizations manufacturing parts designated by the customer as "appearance items," the organization shall provide the following:

a) appropriate resources, including lighting, for

組織は，製品又はサービスの承認が，ISO 9001の8.5.6に従って，初回リリースに引き続く変更の後に遂行されることを確実にしなければならない．

### 8.6.2　レイアウト検査及び機能試験

レイアウト検査，並びに該当する顧客の材料及び性能の技術規格に対する機能検証は，コントロールプランに規定されたとおり，各製品に対して実行されなければならない．その結果は，顧客がレビューのために利用できなければならない．

注記1　レイアウト検査は，設計記録に示される全ての製品寸法を完全に測定することである．

注記2　レイアウト検査の頻度は，顧客によって決定される．

### 8.6.3　外観品目

“外観品目”として顧客に指定された組織の製造部品に対して，組織は，次の事項を提供しなければならない．

a)　照明を含む，評価のための適切な資源

evaluation;

b) masters for colour, grain, gloss, metallic brilliance, texture, distinctness of image (DOI), and haptic technology, as appropriate;
c) maintenance and control of appearance masters and evaluation equipment;
d) verification that personnel making appearance evaluations are competent and qualified to do so.

### 8.6.4 Verification and acceptance of conformity of externally provided products and services

The organization shall have a process to ensure the quality of externally provided processes, products, and services utilizing one or more of the following methods:

a) receipt and evaluation of statistical data provided by the supplier to the organization;
b) receiving inspection and/or testing, such as sampling based on performance;
c) second-party or third-party assessments or audits of supplier sites when coupled with records of acceptable delivered product confor-

b)　必要に応じて，色，絞，光沢，金属性光沢，風合い，イメージの明瞭さ（DOI）のマスター及び触覚技術
c)　外観マスター及び評価設備の保全及び管理

d)　外観評価を実施する要員が力量をもち適格性確認されていることの検証

### 8.6.4　外部から提供される製品及びサービスの検証及び受入れ

組織は，次の方法を一つ以上用いて，外部から提供されるプロセス，製品及びサービスの品質を確実にするプロセスをもたなければならない．

a)　供給者から組織に提供された統計データの受領及び評価
b)　パフォーマンスに基づく抜取検査のような，受入検査及び／又は試験
c)　受入れ可能な納入製品の要求事項への適合の記録を伴う場合，供給者のサイトの第二者若しくは第三者の評価又は監査

mance to requirements;

d) part evaluation by a designated laboratory;

e) another method agreed with the customer.

### 8.6.5 Statutory and regulatory conformity

Prior to release of externally provided products into its production flow, the organization shall confirm and be able to provide evidence that externally provided processes, products, and services conform to the latest applicable statutory, regulatory, and other requirements in the countries where they are manufactured and in the customer-identified countries of destination, if provided.

### 8.6.6 Acceptance criteria

Acceptance criteria shall be defined by the organization and, where appropriate or required, approved by the customer. For attribute data sampling, the acceptance level shall be zero defects (see Section 9.1.1.1).

## 8.7 Control of nonconforming outputs

### 8.7.1

See ISO 9001:2015 requirements.

d)　指定された試験所による部品評価
e)　顧客と合意した他の方法

### 8.6.5　法令・規制への適合

自社の生産フローに外部から提供される製品をリリースする前に，組織は，外部から提供されるプロセス，製品及びサービスが，製造された国及び提供されれば顧客指定の仕向国における最新の法令，規制及び他の要求事項に適合していることを確認し，それを証明する証拠を提供できなければならない．

### 8.6.6　合否判定基準

合否判定基準は，組織によって定められ，必要に応じて又は要求がある場合，顧客に承認されなければならない．抜取りで得られた計数データの合否判定水準は，不良ゼロ（ゼロ・ディフェクト）でなければならない（9.1.1.1 参照）．

## 8.7　不適合なアウトプットの管理

### 8.7.1

ISO 9001:2015 要求事項参照．

### 8.7.1.1 Customer authorization for concession

The organization shall obtain a customer concession or deviation permit prior to further processing whenever the product or manufacturing process is different from that which is currently approved.

The organization shall obtain customer authorization prior to further processing for "use as is" and rework dispositions of nonconforming product. If sub-components are reused in the manufacturing process, that sub-component reuse shall be clearly communicated to the customer in the concession or deviation permit.

The organization shall maintain a record of the expiration date or quantity authorized under concession. The organization shall also ensure compliance with the original or superseding specifications and requirements when the authorization expires. Material shipped under concession shall be properly identified on each shipping container (this applies equally to purchased product). The

#### 8.7.1.1　特別採用に対する顧客の正式許可

組織は，製品又は製造工程が現在承認されているものと異なる場合は常に，その後の処理の前に顧客の特別採用又は逸脱許可を得なければならない.

組織は，不適合製品の“現状での使用”及び手直しをする処置について以降の処理を進める前に顧客の正式許可を受けなければならない．もし構成部品が製造工程で再使用される場合は，その構成部品は，特別採用又は逸脱許可によって，明確に顧客に伝達しなければならない.

組織は，特別採用によって正式許可された満了日又は数量の記録を維持しなければならない．組織は，正式許可が満了となったときには，元の又は置き換わった新たな仕様書及び要求事項に適合していることを確実にしなければならない．特別採用として出荷される材料は，各出荷容器上で適切に識別されなければならない（これは，購入された製品にも等しく適用する.）．組織は供給者からの要請に対し

organization shall approve any requests from suppliers before submission to the customer.

**8.7.1.2 Control of nonconforming product — customer-specified process**

The organization shall comply with applicable customer-specified controls for nonconforming product(s).

**8.7.1.3 Control of suspect product**

The organization shall ensure that product with unidentified or suspect status is classified and controlled as nonconforming product. The organization shall ensure that all appropriate manufacturing personnel receive training for containment of suspect and nonconforming product.

**8.7.1.4 Control of reworked product**

The organization shall utilize risk analysis (such as FMEA) methodology to assess risks in the rework process prior to a decision to rework the product. If required by the customer, the organization shall obtain approval from the customer prior to commencing rework of the product.

て，顧客に提出する前に，承認しなければならない．

#### 8.7.1.2　不適合製品の管理—顧客規定のプロセス

組織は，不適合製品に対して該当する顧客規定の管理に従わなければならない．

#### 8.7.1.3　疑わしい製品の管理

組織は，未確認の又は疑わしい状態の製品を，不適合製品として分類し管理することを確実にしなければならない．組織は，全ての適切な製造要員が，疑わしい製品及び不適合製品の封じ込めの教育訓練を受けることを確実にしなければならない．

#### 8.7.1.4　手直し製品の管理

組織は，製品を手直しする判断の前に，手直し工程におけるリスクを評価するために，リスク分析（FMEAのような）の方法論を活用しなければならない．顧客から要求される場合，組織は，製品の手直しを開始する前に，顧客から承認を取得しなければならない．

The organization shall have a documented process for rework confirmation in accordance with the control plan or other relevant documented information to verify compliance to original specifications.

Instructions for disassembly or rework, including re-inspection and traceability requirements, shall be accessible to and utilized by the appropriate personnel.

The organization shall retain documented information on the disposition of reworked product including quantity, disposition, disposition date, and applicable traceability information.

### 8.7.1.5 Control of repaired product

The organization shall utilize risk analysis (such as FMEA) methodology to assess risks in the repair process prior to a decision to repair the product. The organization shall obtain approval from the customer before commencing repair of the product.

組織は，コントロールプラン又は他の関連する文書化した情報に従って，原仕様への適合を検証する手直し確認の文書化したプロセスをもたなければならない．

再検査及びトレーサビリティ要求事項を含む，分解又は手直し指示書は，適切な要員がアクセスでき，利用できなければならない．

組織は，量，処置，処置日及び該当するトレーサビリティ情報を含めて，手直しした製品の処置に関する文書化した情報を保持しなければならない．

### 8.7.1.5　修理製品の管理

組織は，製品を修理する判断の前に，修理工程におけるリスクを評価するために，リスク分析（FMEAのような）の方法論を活用しなければならない．組織は，製品の修理を開始する前に，顧客から承認を取得しなければならない．

The organization shall have a documented process for repair confirmation in accordance with the control plan or other relevant documented information.

Instructions for disassembly or repair, including re-inspection and traceability requirements, shall be accessible to and utilized by the appropriate personnel.

The organization shall obtain a documented customer authorization for concession for the product to be repaired.

The organization shall retain documented information on the disposition of repaired product including quantity, disposition, disposition date, and applicable traceability information.

**8.7.1.6 Customer notification**

The organization shall immediately notify the customer(s) in the event that nonconforming product has been shipped. Initial communication shall be followed with detailed documentation of the

組織は，コントロールプラン又は他の関連する文書化した情報に従って，修理確認の文書化したプロセスをもたなければならない．

再検査及びトレーサビリティ要求事項を含む，分解又は修理指示書は，適切な要員がアクセスでき，利用できなければならない．

組織は，修理される製品の特別採用について，文書化した顧客の正式許可を取得しなければならない．

組織は，量，処置，処置日及び該当するトレーサビリティ情報を含めて，修理した製品の処置に関する文書化した情報を保持しなければならない．

#### 8.7.1.6　顧客への通知

不適合製品が出荷された場合には，顧客に対して速やかに通知しなければならない．初回の伝達に引き続き，その事象の詳細な文書を提供しなければならない．

event.

#### 8.7.1.7 Nonconforming product disposition

The organization shall have a documented process for disposition of nonconforming product not subject to rework or repair. For product not meeting requirements, the organization shall verify that the product to be scrapped is rendered unusable prior to disposal.

The organization shall not divert nonconforming product to service or other use without prior customer approval.

**8.7.2**

See ISO 9001:2015 requirements.

# 9 Performance evaluation

## 9.1 Monitoring, measurement, analysis and evaluation

### 9.1.1 General

See ISO 9001:2015 requirements.

### 8.7.1.7 不適合製品の廃棄

組織は，手直し又は修理できない不適合製品の廃棄に関する文書化したプロセスをもたなければならない．要求事項を満たさない製品に対して，組織は，スクラップされる製品が廃棄の前に使用不可の状態にされていることを検証しなければならない．

組織は，事前の顧客承認なしで，不適合製品をサービス又は他の使用に流用してはならない．

**8.7.2**

ISO 9001:2015 要求事項参照．

## 9 パフォーマンス評価

### 9.1 監視，測定，分析及び評価

#### 9.1.1 一般

ISO 9001:2015 要求事項参照．

#### 9.1.1.1 Monitoring and measurement of manufacturing processes

The organization shall perform process studies on all new manufacturing (including assembly or sequencing) processes to verify process capability and to provide additional input for process control, including those for special characteristics.

NOTE For some manufacturing processes, it may not be possible to demonstrate product compliance through process capability. For those processes, alternate methods such as batch conformance to specification may be used.

The organization shall maintain manufacturing process capability or performance results as specified by the customer's part approval process requirements. The organization shall verify that the process flow diagram, PFMEA, and control plan are implemented, including adherence to the following:

a) measurement techniques;

### 9.1.1.1　製造工程の監視及び測定

組織は，全ての新規製造工程（組立又は整列を含む．）に対して，工程能力を検証し，特殊特性の管理を含む工程管理への追加インプットを提供するために，工程調査を実施しなければならない．

注記　製造工程によっては，工程能力を通じて製品適合を実証することができない場合がある．それらの工程に対しては，仕様書に対する一括適合のような代替の方法を採用してもよい．

組織は，顧客の部品承認プロセス要求事項で規定された製造工程能力($C_{pk}$)*又は製造工程性能($P_{pk}$)*の結果を維持しなければならない．組織は，工程フロー図，PFMEA及びコントロールプランが実施されることを確実にしなければならない．これには次の事項の順守を含めなければならない．

a)　測定手法

---

*　括弧内の記号は，AIAGの承認のもと，日本語版にのみ加筆したものである．

b) sampling plans;

c) acceptance criteria;

d) records of actual measurement values and/or test results for variable data;

e) reaction plans and escalation process when acceptance criteria are not met.

Significant process events, such as tool change or machine repair, shall be recorded and retained as documented information.

The organization shall initiate a reaction plan indicated on the control plan and evaluated for impact on compliance to specifications for characteristics that are either not statistically capable or are unstable. These reaction plans shall include containment of product and 100 percent inspection, as appropriate. A corrective action plan shall be developed and implemented by the organization indicating specific actions, timing, and assigned responsibilities to ensure that the process becomes stable and statistically capable. The plans shall be reviewed with and approved by the customer, when required.

b)　抜取計画
c)　合否判定基準
d)　変数データに対する実際の測定値及び／又は試験結果の記録
e)　合否判定基準が満たされない場合の対応計画及び上申プロセス

治工具の変更，機械の修理のような工程の重大な出来事は，文書化した情報として記録し保持しなければならない．

組織は，統計的に能力不足又は不安定のいずれかである特性に対して，コントロールプランに記載されている，仕様への適合の影響が評価された対応計画を開始しなければならない．対応計画には，必要に応じて，製品の封じ込め及び全数検査を含めなければならない．工程が安定し，統計的に能力をもつようになることを確実にするために，特定の処置，時期及び担当責任者を規定する是正処置計画を策定し実施しなければならない．この計画は，要求される場合，顧客とともにレビューし承認を得なければならない．

The organization shall maintain records of effective dates of process changes.

#### 9.1.1.2 Identification of statistical tools

The organization shall determine the appropriate use of statistical tools. The organization shall verify that appropriate statistical tools are included as part of the advanced product quality planning (or equivalent) process and included in the design risk analysis (such as DFMEA) (where applicable), the process risk analysis (such as PFMEA), and the control plan.

#### 9.1.1.3 Application of statistical concepts

Statistical concepts, such as variation, control (stability), process capability, and the consequences of over-adjustment, shall be understood and used by employees involved in the collection, analysis, and management of statistical data.

### 9.1.2 Customer satisfaction

See ISO 9001:2015 requirements.

組織は，工程変更の実効日付の記録を維持しなければならない．

### 9.1.1.2　統計的ツールの特定

組織は，統計的ツールの適切な使い方を決定しなければならない．組織は，適切な統計的ツールが先行製品品質計画（又はそれに相当する）プロセスの一部として含まれていること，かつ，設計リスク分析（DFMEA のような）（該当する場合には，必ず），工程リスク分析（PFMEA のような）及びコントロールプランに含まれていることを検証しなければならない．

### 9.1.1.3　統計概念の適用

ばらつき，管理（安定性），工程能力及び過剰調整によって起きてしまう結果のような統計概念は，統計データの収集，分析及び管理に携わる従業員に理解され，使用されなければならない．

### 9.1.2　顧客満足

ISO 9001:2015 要求事項参照．

### 9.1.2.1 Customer satisfaction — supplemental

Customer satisfaction with the organization shall be monitored through continual evaluation of internal and external performance indicators to ensure compliance to the product and process specifications and other customer requirements.

Performance indicators shall be based on objective evidence and include but not be limited to the following:

a) delivered part quality performance;
b) customer disruptions;
c) field returns, recalls, and warranty (where applicable);
d) delivery schedule performance (including incidents of premium freight);
e) customer notifications related to quality or delivery issues, including special status.

The organization shall monitor the performance of manufacturing processes to demonstrate compliance with customer requirements for product quality and process efficiency. The monitoring

### 9.1.2.1　顧客満足―補足

製品及びプロセスの仕様書及び他の顧客要求事項への適合を確実にするために，内部及び外部の評価指標の継続的評価を通じて，組織に対する顧客満足を監視しなければならない．

パフォーマンス指標は，客観的証拠に基づき，次の事項を含めなければならない．しかし，それに限定されない．

a)　納入した部品の品質パフォーマンス
b)　顧客が被った迷惑
c)　市場で起きた回収，リコール，補償（該当する場合には，必ず）
d)　納期パフォーマンス（特別輸送費が発生する不具合を含む．）
e)　品質又は納期問題に関する，特別状態を含む，顧客からの通知

組織は，製品品質及びプロセス効率に対する顧客要求事項への適合を実証するために，製造工程のパフォーマンスを監視しなければならない．監視には，提供される場合，オンライン顧客ポータル及び

shall include the review of customer performance data including online customer portals and customer scorecards, where provided.

### 9.1.3 Analysis and evaluation

See ISO 9001:2015 requirements.

### 9.1.3.1 Prioritization

Trends in quality and operational performance shall be compared with progress toward objectives and lead to action to support prioritization of actions for improving customer satisfaction.

## 9.2 Internal audit

### 9.2.1 and 9.2.2

See ISO 9001:2015 requirements.

### 9.2.2.1 Internal audit programme

The organization shall have a documented internal audit process. The process shall include the development and implementation of an internal audit programme that covers the entire quality management system including quality management system audits, manufacturing process au-

顧客スコアカードを含む，顧客パフォーマンスデータのレビューを含めなければならない．

**9.1.3　分析及び評価**

ISO 9001:2015 要求事項参照．

### 9.1.3.1　優先順位付け

品質及び運用パフォーマンスの傾向は，目標への進展と比較し，顧客満足を改善する処置の優先順位付けを支援する処置を導かなければならない．

**9.2　内部監査**

**9.2.1 及び 9.2.2**

ISO 9001:2015 要求事項参照．

### 9.2.2.1　内部監査プログラム

組織は，文書化した内部監査プロセスをもたなければならない．そのプロセスには，品質マネジメントシステム監査，製造工程監査及び製品監査を含む，品質マネジメントシステム全体を網羅する内部監査プログラムの策定及び実施を含めなければならない．

dits, and product audits.

The audit programme shall be prioritized based upon risk, internal and external performance trends, and criticality of the process(es).

Where the organization is responsible for software development, the organization shall include software development capability assessments in their internal audit programme.

The frequency of audits shall be reviewed and, where appropriate, adjusted based on occurrence of process changes, internal and external nonconformities, and/or customer complaints. The effectiveness of the audit programme shall be reviewed as a part of management review.

**9.2.2.2 Quality management system audit**

The organization shall audit all quality management system processes over each three-year calendar period, according to an annual programme, using the process approach to verify compliance with this Automotive QMS Standard. Integrated

監査プログラムは，リスク，内部及び外部パフォーマンスの傾向及びプロセスの重大性に基づいて優先付けしなければならない．

組織がソフトウェア開発の責任がある場合，組織は，ソフトウェア開発能力評価を監査プログラムに含めなければならない．

プロセス変更，内部及び外部不適合，並びに顧客苦情に基づいて，監査頻度をレビューし，必要に応じて，調整しなければならない．監査プログラムの有効性は，マネジメントレビューの一部としてレビューしなければならない．

### 9.2.2.2　品質マネジメントシステム監査

組織は，この自動車産業 QMS 規格への適合を検証するために，プロセスアプローチを使用して，各3暦年の期間，年次プログラムに従って，全ての品質マネジメントシステムのプロセスを監査しなければならない．それらの監査に統合させて，組織は，

with these audits, the organization shall sample customer-specific quality management system requirements for effective implementation.

#### 9.2.2.3 Manufacturing process audit

The organization shall audit all manufacturing processes over each three-year calendar period to determine their effectiveness and efficiency using customer-specific required approaches for process audits. Where not defined by the customer, the organization shall determine the approach to be used.

Within each individual audit plan, each manufacturing process shall be audited on all shifts where it occurs, including the appropriate sampling of the shift handover.

The manufacturing process audit shall include an audit of the effective implementation of the process risk analysis (such as PFMEA), control plan, and associated documents.

#### 9.2.2.4 Product audit

顧客固有の品質マネジメントシステム要求事項が，効果的に実施されているかをサンプリング確認しなければならない．

### 9.2.2.3　製造工程監査

組織は，製造工程の有効性及び効率を判定するために，各3暦年の期間，工程監査のための顧客固有に要求される方法を使用して，全ての製造工程を監査しなければならない．顧客によって定められていない場合，組織は，使用する方法を決めなければならない．

個別の各監査計画の中で，各製造工程は，シフト引継ぎの適切なサンプリングを含めて，それが行われている全てのシフトを監査しなければならない．

製造工程監査には，工程リスク分析（PFMEAのような），コントロールプラン及び関連文書が効果的に実施されているかの監査を含めなければならない．

### 9.2.2.4　製品監査

The organization shall audit products using customer-specific required approaches at appropriate stages of production and delivery to verify conformity to specified requirements. Where not defined by the customer, the organization shall define the approach to be used.

## 9.3 Management review

### 9.3.1 General

See ISO 9001:2015 requirements.

#### 9.3.1.1 Management review — supplemental

Management review shall be conducted at least annually. The frequency of management review(s) shall be increased based on risk to compliance with customer requirements resulting from internal or external changes impacting the quality management system and performance-related issues.

### 9.3.2 Management review inputs

See ISO 9001:2015 requirements.

組織は，規定要求事項への適合を検証するために，顧客固有の要求される方法を使用して，生産及び引渡しの適切な段階で，製品を監査しなければならない．顧客によって定められていない場合，組織は，使用する方法を定めなければならない．

## 9.3　マネジメントレビュー

### 9.3.1　一般

ISO 9001:2015 要求事項参照．

#### 9.3.1.1　マネジメントレビュー—補足

マネジメントレビューは，少なくとも年次で実施しなければならない．品質マネジメントシステム及びパフォーマンスに関係する問題に影響する内部又は外部の変化による顧客要求事項への適合のリスクに基づいて，マネジメントレビューの頻度を増やさなければならない．

### 9.3.2　マネジメントレビューへのインプット

ISO 9001:2015 要求事項参照．

### 9.3.2.1 Management review inputs — supplemental

Input to management review shall include:

a) cost of poor quality (cost of internal and external nonconformance);

b) measures of process effectiveness;

c) measures of process efficiency;

d) product conformance;

e) assessments of manufacturing feasibility made for changes to existing operations and for new facilities or new product (see Section 7.1.3.1);

f) customer satisfaction (see ISO 9001, Section 9.1.2);

g) review of performance against maintenance objectives;

h) warranty performance (where applicable);

i) review of customer scorecards (where applicable);

j) identification of potential field failures identified through risk analysis (such as FMEA);

k) actual field failures and their impact on safe-

### 9.3.2.1 マネジメントレビューへのインプット―補足

マネジメントレビューへのインプットには，次の事項を含めなければならない．

a) 品質不良コスト（内部不適合及び外部不適合のコスト）
b) プロセスの有効性の対策
c) プロセスの効率の対策
d) 製品適合性
e) 現行の運用の変更及び新規施設又は新規製品に対してなされる製造フィージビリティ評価（7.1.3.1 参照）
f) 顧客満足（ISO 9001 の 9.1.2 参照）
g) 保全目標に対するパフォーマンスのレビュー
h) 補償のパフォーマンス（該当する場合には，必ず）
i) 顧客スコアカードのレビュー（該当する場合には，必ず）
j) リスク分析（FMEA のような）を通じて明確にされた潜在的市場不具合の特定
k) 実際の市場不具合及びそれらが安全又は環境に

ty or the environment.

### 9.3.3 Management review outputs

See ISO 9001:2015 requirements.

### 9.3.3.1 Management review outputs — supplemental

Top management shall document and implement an action plan when customer performance targets are not met.

# 10 Improvement

## 10.1 General

See ISO 9001:2015 requirements.

## 10.2 Nonconformity and corrective action

### 10.2.1 and 10.2.2

See ISO 9001:2015 requirements.

### 10.2.3 Problem solving

The organization shall have a documented process(es) for problem solving including:

a) defined approaches for various types and scale of problems (e.g., new product develop-

与える影響

**9.3.3　マネジメントレビューからのアウトプット**

ISO 9001:2015 要求事項参照.

**9.3.3.1　マネジメントレビューからのアウトプット—補足**

トップマネジメントは，顧客のパフォーマンス目標が達成されていない場合には，処置計画を文書化し実施しなければならない.

**10　改善**

**10.1　一般**

ISO 9001:2015 要求事項参照.

**10.2　不適合及び是正処置**

**10.2.1 及び 10.2.2**

ISO 9001:2015 要求事項参照.

**10.2.3　問題解決**

組織は，次の事項を含む，問題解決の方法を文書化したプロセスをもたなければならない.

a)　問題の様々なタイプ及び規模に対する定められたアプローチの仕方（例　新製品開発，現行製

ment, current manufacturing issues, field failures, audit findings);

b) containment, interim actions, and related activities necessary for control of nonconforming outputs (see ISO 9001, Section 8.7);

c) root cause analysis, methodology used, analysis, and results;

d) implementation of systemic corrective actions, including consideration of the impact on similar processes and products;

e) verification of the effectiveness of implemented corrective actions;

f) reviewing and, where necessary, updating the appropriate documented information (e.g., PFMEA, control plan).

Where the customer has specific prescribed processes, tools, or systems for problem solving, the organization shall use those processes, tools, or systems unless otherwise approved by the customer.

### 10.2.4 Error-proofing

The organization shall have a documented process

造問題，市場不具合，監査所見）

b)　不適合なアウトプット（ISO 9001 の 8.7 参照）の管理に必要な，封じ込め，暫定処置及び関係する活動

c)　根本原因分析，使用される方法論，分析及び結果

d)　類似のプロセス及び製品への影響を考慮することを含む，体系的是正処置の実施

e)　実施された是正処置の有効性検証

f)　適切な文書化した情報（例　PFMEA，コントロールプラン）のレビュー及び，必要に応じた更新

顧客が固有の規定されたプロセス，ツール，又は問題解決のシステムをもっている場合，顧客によって他に承認がない限り，組織は，そのプロセス，ツール，又はシステムを使用しなければならない．

### 10.2.4　ポカヨケ

組織は，適切なポカヨケ手法の活用について決

to determine the use of appropriate error-proofing methodologies. Details of the method used shall be documented in the process risk analysis (such as PFMEA) and test frequencies shall be documented in the control plan.

The process shall include the testing of error-proofing devices for failure or simulated failure. Records shall be maintained. Challenge parts, when used, shall be identified, controlled, verified, and calibrated where feasible. Error-proofing device failures shall have a reaction plan.

### 10.2.5 Warranty management systems

When the organization is required to provide warranty for their product(s), the organization shall implement a warranty management process. The organization shall include in the process a method for warranty part analysis, including NTF (no trouble found). When specified by the customer, the organization shall implement the required warranty management process.

定する文書化したプロセスをもたなければならない．採用された手法の詳細は，プロセスリスク分析(PFMEA のような）に文書化し，試験頻度はコントロールプランに文書化しなければならない．

そのプロセスには，ポカヨケ装置の故障又は模擬故障のテストを含めなければならない．記録は維持しなければならない．チャレンジ部品が使用される場合，実現可能であれば，識別し，管理し，検証し，及び校正しなければならない．ポカヨケ装置の故障には，対応計画をもたなければならない．

### 10.2.5　補償管理システム

組織の製品に対して補償を要求される場合，組織は，補償管理プロセスを実施しなければならない．組織は，NTF（no trouble found）を含めて，そのプロセスに補償部品分析の方法を含めなければならない．顧客に規定されている場合，組織は，その要求される補償管理プロセスを実施しなければならない．

#### 10.2.6 Customer complaints and field failure test analysis

The organization shall perform analysis on customer complaints and field failures, including any returned parts, and shall initiate problem solving and corrective action to prevent recurrence.

Where requested by the customer, this shall include analysis of the interaction of embedded software of the organization's product within the system of the final customer's product.

The organization shall communicate the results of testing/analysis to the customer and also within the organization.

### 10.3 Continual improvement

See ISO 9001:2015 requirements.

#### 10.3.1 Continual improvement — supplemental

The organization shall have a documented process for continual improvement. The organization shall include in this process the following:

### 10.2.6　顧客苦情及び市場不具合の試験・分析

組織は，顧客苦情及び市場不具合に対して，回収された部品を含めて，分析しなければならない．そして，再発防止のために問題解決及び是正処置を開始しなければならない．

顧客に要求された場合は，これには，顧客の最終製品内での，組織の製品の組込みソフトウェアの相互作用の分析を含めなければならない．

組織は，試験／分析の結果を，顧客及び組織内にも伝達しなければならない．

## 10.3　継続的改善

ISO 9001:2015 要求事項参照．

### 10.3.1　継続的改善—補足

組織は，継続的改善の文書化したプロセスをもたなければならない．組織は，このプロセスに次の事項を含めなければならない．

a) identification of the methodology used, objectives, measurement, effectiveness, and documented information;

b) a manufacturing process improvement action plan with emphasis on the reduction of process variation and waste;

c) risk analysis (such as FMEA).

NOTE Continual improvement is implemented once manufacturing processes are statistically capable and stable or when product characteristics are predictable and meet customer requirements.

a)　使用される方法論，目標，評価指標，有効性及び文書化した情報の明確化

b)　工程ばらつき及び無駄の削減に重点を置いた，製造工程の改善計画

c)　リスク分析（FMEA のような）

注記　継続的改善は，製造工程が統計的に能力をもち安定してから，又は製品特性が予測可能で顧客要求事項を満たしてから，実施される.

# Annex A: Control Plan

## A.1 Phases of the control plan

A control plan covers three distinct phases, as appropriate:

a) **Prototype**: a description of the dimensional measurements, material, and performance tests that will occur during building of the prototype. The organization shall have a prototype control plan, if required by the customer.

b) **Pre-launch**: a description of the dimensional measurements, material, and performance tests that occur after prototype and before full production. Pre-launch is defined as a production phase in the process of product realization that may be required after prototype build.

c) **Production**: documentation of product/process characteristics, process controls, tests, and measurement systems that occur during mass production.

Control plans are established at a part number

# 附属書 A：コントロールプラン

## A.1　コントロールプランの段階

コントロールプランは，必要に応じて，三つの区分された段階を網羅する．

a)　**試作**：試作中に行われる寸法測定，材料及び性能試験を記述したもの．組織は，顧客に要求される場合，試作コントロールプランを作成しなければならない．

b)　**量産試作**：試作後で量産の前に行われる寸法測定，材料及び性能試験を記述したもの．量産試作は，試作後に要求されることがある製品実現プロセスにおける量産段階の一つとして定義されている．

c)　**量産**：量産中に行われる，製品／工程の特性，工程管理，試験及び測定システムを記述した文書．

個々の部番・品番ごとにコントロールプランを作

level; but in many cases, family control plans may cover a number of similar parts produced using a common process. Control plans are an output of the quality plan.

NOTE 1 It is recommended that the organization require its suppliers to meet the requirements of this Annex.

NOTE 2 For some bulk materials, the control plans do not list most of the production information. This information can be found in the corresponding batch formulation/recipe details.

## A.2 Elements of the control plan

A control plan includes, as a minimum, the following contents:

### General data

a) control plan number

b) issue date and revision date, if any

c) customer information (see customer requirements)

d) organization's name/site designation

e) part number(s)

成しなければならない．しかし，多くの場合，ファミリーコントロールプランは，共通の工程を用いて生産する同種の部品群を網羅する場合がある．コントロールプランは品質計画のアウトプットである．

注記1　組織は，その供給者に，この附属書の要求事項を満たすように要求することを推奨する．

注記2　バルク材料によっては，コントロールプランには生産情報の多くは記載されない．この情報は，該当バッチ処方／レシピ詳細に見いだせる．

## A.2　コントロールプランの要素

コントロールプランには，最低限，次の内容を含める．

**一般データ**

a）　コントロールプラン番号

b）　発行日付及びもしあれば改訂日付

c）　顧客情報（顧客要求事項参照）

d）　組織名称／サイト名称

e）　部品番号

f) part name/description
g) engineering change level
h) phase covered (prototype, pre-launch, production)
i) key contact
j) part/process step number
k) process name/operation description
l) functional group/area responsible

**Product control**

a) product-related special characteristics
b) other characteristics for control (number, product or process)
c) specification/tolerance

**Process control**

a) process parameters (including process settings and tolerances)
b) process-related special characteristics
c) machines, jigs, fixtures, tools for manufacturing (including identifiers, as appropriate)

**Methods**

a) evaluation measurement technique

f)　部品名称／摘要

g)　技術変更レベル

h)　対応段階（試作，量産試作，量産）

i)　主要連絡先

j)　部品／工程ステップ番号

k)　工程名称／作業説明

l)　機能グループ／責任領域

**製品管理**

a)　製品に関する特殊特性

b)　管理のための他の特性（番号，製品又は工程）

c)　仕様／公差

**工程管理**

a)　工程パラメータ（工程設定及び公差を含む.）

b)　工程に関する特殊特性

c)　製造のための機械，ジグ，固定装置，工具（必要に応じて識別子を含む.）

**方法**

a)　評価測定技法

b) error-proofing

c) sample size and frequency

d) control method

## Reaction plan

a) reaction plan (include or reference)

b） ポカヨケ

c） 抜取り数及び頻度

d） 管理方法

**対応計画**

a） 対応計画（含む又は参照する.）

# ANNEX B: Bibliography — supplemental automotive

## Internal audit

AIAG

CQI-8 Layered Process Audit

CQI-9 Special Process: Heat Treatment System Assessment

CQI-11 Special Process: Plating System Assessment

CQI-12 Special Process: Coating System Assessment

CQI-15 Special Process: Welding System Assessment

CQI-17 Special Process: Soldering System Assessment

CQI-23 Special Process: Molding System Assessment

CQI-27 Special Process: Casting System Assessment

ANFIA

AQ 008 Process Audit

FIEV

V2.0 Production Process Audit Manual

# 附属書 B：参考文献
# —自動車産業補足

## 内部監査

AIAG

CQI-8　Layered Process Audit

CQI-9　Special Process: Heat Treatment System Assessment

CQI-11　Special Process: Plating System Assessment

CQI-12　Special Process: Coating System Assessment

CQI-15　Special Process: Welding System Assessment

CQI-17　Special Process: Soldering System Assessment

CQI-23　Special Process: Molding System Assessment

CQI-27　Special Process: Casting System Assessment

ANFIA

AQ 008 Process Audit

FIEV

V2.0 Production Process Audit Manual

IATF

Auditor Guide for IATF 16949

VDA

Volume 6 part 3 Process Audit

Volume 6 part 5 Product Audit

**Nonconformity and corrective action**

AIAG

CQI-14 Automotive Warranty Management Guideline

CQI-20 Effective Problem Solving Practitioner Guide

VDA

Volume "Audit standard field failure analysis"

Volume "Field failures analysis"

**Measurement system analysis**

AIAG

Measurement Systems Analysis (MSA)

ANFIA

AQ 024 MSA Measurement Systems Analysis

VDA

Volume 5 "Capability of Measuring Systems"

IATF

Auditor Guide for IATF 16949

VDA

Volume 6 part 3 Process Audit

Volume 6 part 5 Product Audit

### 不適合及び是正処置

AIAG

CQI-14 Automotive Warranty Management Guideline

CQI-20 Effective Problem Solving Practitioner Guide

VDA

Volume “Audit standard field failure analysis”

Volume “Field failures analysis”

### 測定システム解析

AIAG

Measurement Systems Analysis (MSA)

ANFIA

AQ 024 MSA Measurement Systems Analysis

VDA

Volume 5 “Capability of Measuring Systems”

**Product approval**

AIAG

Production Part Approval Process (PPAP)

VDA

Volume 2 Production process and product approval (PPA)

Volume 19 Part 1 ("Inspection of Technical Cleanliness — Particulate Contamination of Functionally Relevant Automotive Components")

Volume 19 Part 2 ("Technical cleanliness in assembly — Environment, Logistics, Personnel and Assembly Equipment")

**Product design**

AIAG

APQP and Control Plan

CQI-24 Design Review Based on Failure Modes (DRBFM Reference Guide)

Potential Failure Mode & Effects Analysis (FMEA)

ANFIA

AQ 009 FMEA

## 製品承認

AIAG

Production Part Approval Process (PPAP)

VDA

Volume 2 Production process and product approval (PPA)

Volume 19 Part 1 (“Inspection of Technical Cleanliness — Particulate Contamination of Functionally Relevant Automotive Components”)

Volume 19 Part 2 (“Technical cleanliness in assembly — Environment, Logistics, Personnel and Assembly Equipment”)

## 製品設計

AIAG

APQP and Control Plan

CQI-24 Design Review Based on Failure Modes (DRBFM Reference Guide)

Potential Failure Mode & Effects Analysis (FMEA)

ANFIA

AQ 009 FMEA

AQ 014 Manual of Experimental Design

AQ 025 Reliability Guide

VDA

Volume 4 Chapter Product and Process FMEA

Volume VDA-RGA "Maturity Level Assurance for New Parts"

Volume "Robust Production Process"

Volume Special Characteristics (SC)

**Production control**

AIAG

MMOG/LE Materials Management Operational Guidelines / Logistics Evaluation

SMMT

Implementing Standardised Work

**Quality management system administration**

ANFIA

AQ 026 Managing and improving the process

IATF

Rules for achieving and maintaining IATF recognition

AQ 014 Manual of Experimental Design

AQ 025 Reliability Guide

VDA

Volume 4 Chapter Product and Process FMEA

Volume VDA-RGA "Maturity Level Assurance for New Parts"

Volume "Robust Production Process"

Volume Special Characteristics (SC)

**生産管理**

AIAG

MMOG/LE Materials Management Operational Guidelines / Logistics Evaluation

SMMT

Implementing Standardised Work

**品質マネジメントシステム運営**

ANFIA

AQ 026 Managing and improving the process

IATF

Rules for achieving and maintaining IATF recognition

**Risk analysis**

VDA

Volume 4 "Ring-binder" (elementary aids, risk analyses, methods, and process models)

**Software Process Assessment**

Capability Maturity Model Integration (CMMI)

VDA

Automotive SPICE® (Software Process Improvement and Capability Determination)

**Statistical tools**

AIAG

Statistical Process Control (SPC)

ANFIA

AQ 011 SPC

**Supplier quality management**

AIAG

CQI-19 Sub-Tier Supplier Management Process Guideline

IATF

**リスク分析**

VDA

Volume 4 "Ring-binder" (elementary aids, risk analyses, methods, and process models)

**ソフトウェアプロセス評価**

Capability Maturity Model Integration (CMMI)

VDA

Automotive SPICE® (Software Process Improvement and Capability Determination)

**統計的ツール**

AIAG

Statistical Process Control (SPC)

ANFIA

AQ 011 SPC

**供給者の品質マネジメント**

AIAG

CQI-19 Sub-Tier Supplier Management Process Guideline

IATF

Minimum Automotive Quality Management System Requirements for Sub-Tier Suppliers (MAQMSR)

## Health and safety

ISO

ISO 45001 Occupational health and safety management systems

Minimum Automotive Quality Management System Requirements for Sub-Tier Suppliers (MAQMSR)

## 安全衛生

ISO

ISO 45001 Occupational health and safety management systems

**対訳 IATF 16949:2016**
**自動車産業品質マネジメントシステム規格**
**—自動車産業の生産部品及び関連するサービス部品の**
**組織に対する品質マネジメントシステム要求事項**
**［ポケット版］**

2016 年 12 月 15 日　第 1 版第 1 刷発行
2025 年 1 月 17 日　第 11 刷発行

編　者　一般財団法人 日本規格協会

発 行 者　朝日　弘

発 行 所　一般財団法人 日本規格協会
〒 108–0073　東京都港区三田 3 丁目 11–28 三田 Avanti
https://www.jsa.or.jp/
振替　00160–2–195146

製　作　日本規格協会ソリューションズ株式会社
印 刷 所　三美印刷株式会社


ISBN978–4–542–40276–8